M.

WORK OUT
SERIES

Work Out

Chemistry

GCSE

The titles
in this
series

MACMILLAN
WORK OUT
SERIES

Work Out

Chemistry

GCSE

A. L. Barker
and
K. A. Knapp

MACMILLAN

First published 1986
Reprinted 1986
Reprinted with corrections 1986
This edition 1987

Published by
MACMILLAN EDUCATION LTD
Houndmills, Basingstoke, Hampshire RG21 2XS
and London
Companies and representatives
throughout the world

Typeset by TecSet Ltd,
Wallington, Surrey
Printed and bound in Great Britain by Scotprint Ltd., Musselburgh, Scotland

British Library Cataloguing in Publication Data
Barker, A. L. (Alan Lewis)
Work out chemistry GCSE.—2nd ed.—
(Macmillan work out series)
1. Chemistry
I. Title   II. Knapp, K. A.
540   QD33
ISBN 0-333-44011-0
ISBN 0-333-43451-X (Expt)

Contents

Acknowledgements

The authors and publishers wish to thank the following who have kindly given permission for the use of copyright material:

The Associated Examining Board, the Northern Ireland Schools Examination Council, the Scottish Examination Board, the Southern Universities' Joint Board, the University of London School Examinations Board and the University of Oxford Delegacy of Local Examinations for questions from past examination papers.

Every effort has been made to trace all the copyright holders, but if any have been inadvertently overlooked the publishers will be pleased to make the necessary arrangement at the first opportunity.

The University of London Entrance and School Examinations Council accepts no responsibility whatsoever for the accuracy or method in the answers given in this book to actual questions set by the London Board.

Acknowledgement is made to the Southern Universities' Joint Board for School Examinations for permission to use questions taken from their past papers but the Board is in no way responsible for answers that may be provided and they are solely the responsibility of the authors.

The Associated Examining Board, the University of Oxford Delegacy of Local Examinations, the Northern Ireland Schools Examination Council, and the Scottish Examination Board wish to point out that worked examples included in the text are entirely the responsibility of the authors and have neither been provided nor approved by the Board.

Organisations Responsible for GCSE Examinations

In the United Kingdom, examinations are administered by the following organisations. Syllabuses and examination papers can be ordered from the addresses given here.

Northern Examining Association (NEA)

Joint Matriculation Board (JMB)
Publications available from:
John Sherratt & Son Ltd
78 Park Road, Altrincham
Cheshire WA14 5QQ

North Regional Examinations Board
Wheatfield Road
Westerhope
Newcastle upon Tyne NE5 5JZ

Yorkshire and Humberside Regional Examinations Board (YREB)
Scarsdale House, 136 Derbyside Lane
Sheffield S8 8SE

Associated Lancashire Schools Examining Board
12 Harter Street
Manchester M1 6HL

North West Regional Examinations Board (NWREB)
Orbit House, Albert Street
Eccles, Manchester M30 0WL

Midland Examining Group (MEG)

**University of Cambridge Local
    Examinations Syndicate (UCLES)**
Syndicate Buildings, Hills Road
Cambridge CB1 2EU

**Oxford and Cambridge Schools
    Examination Board (O & C)**
10 Trumpington Street
Cambridge CB2 1QB

Southern Universities' Joint Board (SUJB)
Cotham Road
Bristol BS6 6DD

**East Midland Regional Examinations
    Board (EMREB)**
Robins Wood House, Robins Wood Road
Aspley, Nottingham NG8 3NR

West Midlands Examinations Board (WMEB)
Norfolk House, Smallbrook
Queensway, Birmingham B5 4NJ

London and East Anglian Group (LEAG)

**University of London School
    Examinations Board (L)**
University of London Publications Office
52 Gordon Square
London WC1E 6EE

**London Regional Examining Board
    (LREB)**
Lyon House
104 Wandsworth High Street
London SW18 4LF

East Anglian Examinations Board (EAEB)
The Lindens, Lexden Road
Colchester, Essex CO3 3RL

Southern Examining Group (SEG)

The Associated Examining Board (AEB)
Stag Hill House
Guildford
Surrey GU2 5XJ

**University of Oxford Delegacy of
    Local Examinations (OLE)**
Ewert Place, Banbury Road
Summertown, Oxford OX2 7BZ

**Southern Regional Examinations
    Board (SREB)**
Avondale House
33 Carlton Crescent
Southampton Hants SO9 4YL

**South-East Regional Examinations
    Board (SEREB)**
Beloe House
2–10 Mount Ephraim Road
Royal Tunbridge Wells, Kent TN1 1EU

Scottish Examination Board (SEB)

Publications available from:
Robert Gibson and Sons (Glasgow) Ltd
17 Fitzroy Place
Glasgow G3 7SF

Welsh Joint Education Committee (WJEC)

245 Western Avenue
Cardiff CF5 2YX

Northern Ireland Schools Examinations Council (NISEC)

Examinations Office
Beechill House, Beechill Road
Belfast BT8 4RS

Introduction

How to Use this Book

This book is intended for use towards the end of a Chemistry course leading to an examination for GCSE. Its purpose is to help you with your revision and to improve your examination performance.

Each chapter begins with a short revision section which deals briefly with the topics that you will need to know. For more detailed background reading you should use a textbook such as the authors' *Chemistry — A Practical Approach*.

The revision section is followed by worked examples. These enlarge upon the topics presented in the revision section and help you to see how to present your answers. Further explanation and advice are given occasionally and enclosed in 'boxes' to separate them from the answers.

Finally, there is a section of questions on which you can practise your skills and test your knowledge. Answers are given at the end of each chapter but you should not look at them until you have *written* your own.

If you work carefully and sensibly through the three sections of each chapter you should gain both knowledge and understanding of chemistry and an awareness of what your examiner will require you to do.

Revision

The first thing you must do is to obtain a copy of the syllabus and some specimen papers from your Examination Board so that you know exactly what you have to learn. Many topics are common to the syllabuses of all of the Boards but there is no point at this stage in your course in learning things that you will not be asked about. The addresses of the Examination Boards are given on pages ix–xi.

Next, you should make a revision plan to cover the last few months before the examination. Having made the plan, you must keep to it!

Your revision should be active — try to write out definitions, equations and key points from memory. Every 20 minutes or so, get up and stretch: this may sound silly but it really does help you to concentrate and revise more effectively. Shorten your notes to a few headings, definitions and equations for each topic and learn them thoroughly. There is no excuse for losing marks through not knowing a definition.

Work through questions from the specimen papers in the same way as you do for the self-test sections of this book. You may well find that extra practice, using the authors' *Structured Questions in Chemistry*, will be helpful at this stage.

Remember that advance planning is always better than last-minute cramming.

The Examination

Pack everything you need on the night before the examination and avoid the temptation to stay up late trying to revise. Get up early enough the next morning to enable you to get to the examination room without rushing and becoming flustered.

If the examiner gives you a choice of questions, read the paper carefully before deciding which ones to answer – you may regret it if you choose in a hurry. In all papers, keep an eye on the time and do not spend too long on a difficult question. It is better to leave it and try again later. If you finish early, check your paper carefully — there is always room for improvement.

Types of Question

The Examination Boards use various types of question. You must check to find which ones your Board uses.

Multiple choice questions require you to choose one correct answer from four or five alternatives. Take your time and read **all** of the possible answers given before making your choice. It is very easy to make mistakes in these questions, either by not reading all of the alternatives or by misreading one of them. If you are not sure of the correct answer, you can usually eliminate two or three that are definitely wrong. You can then make a guess from the remainder with a better chance of being right. *Never* leave an answer blank.

In any papers containing this type of question you should first answer the questions which you find easy. Then go back and tackle the more difficult ones if you have time.

Structured questions consist of a series of questions, usually related to one another and requiring short answers (see Example 2.7). Clues to the length of answer and the detail required can be found in the amount of space allowed on the paper for you to write and by the number of marks allocated, which is quoted in a bracket. For example, if 5 marks are to be awarded, a one-word answer is bound to be insufficient.

Free response questions require longer answers (see Example 4.1) and it is worth making a check-list of main points and an outline plan before beginning.

In the last two types of question it is especially important to keep to a time schedule. Draw diagrams only if they are asked for or if they really clarify a point. Draw them neatly but do not spend too long on them as they are rarely worth many marks. Give equations wherever possible because they often earn marks, even when they are not specifically mentioned in the question.

Always answer the question which is asked! This may sound obvious advice but many candidates throw away marks through not paying attention to the wording of the question. If you are asked to *name* a substance, then giving its formula will earn you nothing. If you are asked to *describe* a reaction, the examiner will expect you to mention colour change, fizzing, production of heat, etc. Finally, do not 'waffle' — in other words do not write a great deal on a topic just because you know it well. The examiner will award you marks only if you write what has been asked for.

School-based Assessment of Practical Skills

The assessment of practical skills forms 20% of the GCSE examination. It will be carried out by your teacher over a period of several months. Your teacher will train you to work carefully and efficiently, but it is worth remembering the following points. You must follow exactly all the instructions and use the apparatus and chemicals with care. You must make measurements accurately and record *all* observations. For example, if you are asked to heat copper(II) sulphate crystals and describe what happens, it is not sufficient to say, 'The crystals got hot and a gas was given off.' A better answer would be, 'The blue crystals turn to a white powder; a steamy vapour is given off and this condenses to a colourless liquid at the top of the tube.' Parts of the practical course require you to plan experiments. In order to do this you are going to have to think carefully about what you are trying to achieve. Thinking about experiments you have done in the past may help you to decide on an appropriate method.

Passing examinations requires time and effort. Take note of the advice given by your teachers and by this book. Be prepared to make sacrifices, work steadily, and you should be rewarded by success.

We wish you luck!

A.L.B.
K.A.K.

1 Some Simple Ideas

1.1   Elements, Compounds and Mixtures

An **element** is a substance which cannot be split up into two or more simpler substances by chemical means, e.g. aluminium and sulphur (see Question 1.1).

There are over 100 elements which are grouped into two main classes — metals and non-metals. The differences between the classes are investigated in later chapters (13–17) and a summary appears in Example 13.4.

A **compound** is a substance which consists of two or more elements chemically combined together, e.g. copper(II) sulphate crystals, $CuSO_4.5H_2O$, contain copper, sulphur, oxygen and hydrogen.

Synthesis is the building up of a compound from simpler substances, often its elements.

A **mixture** consists of two or more elements or compounds which have not been chemically combined, e.g. water and ethanol, aluminium and sulphur.

The main differences between compounds and mixtures are summarised in Table 1.1.

Table 1.1

	Compound	Mixture
Composition	Fixed	Variable
Heat change	Heat usually produced or absorbed when compound made	Usually no heat change on making mixture
Appearance	Different from constituent elements	'Average' of constituents
Properties	Different from constituent elements	Similar to constituents
Separation	Cannot be separated into constituent elements by physical means	Can be separated into constituents by physical means

The acronym 'CHAPS' should help you to remember the table.

(a) Abundance of the Elements in the Earth's Crust and Atmosphere (Fig. 1.1)

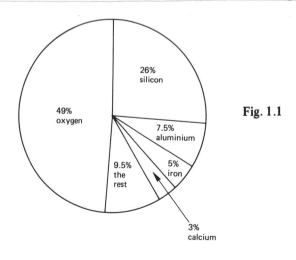

Fig. 1.1

Oxygen is found in the air, combined with hydrogen in water and combined with other elements (e.g. silicon and aluminium) in rocks and soils.

(b) Abundance of Elements in the Human Body (Fig. 1.2)

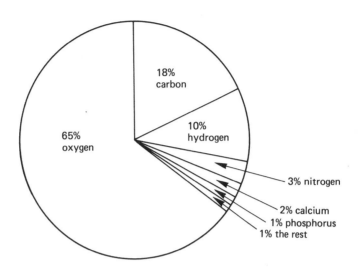

Fig. 1.2

Oxygen, carbon and hydrogen are the main elements in fats, proteins and carbohydrates. Three-quarters of our body weight is water, which accounts for more hydrogen and oxygen. Nitrogen is found in proteins, whereas calcium, together with phosphorus and oxygen, is found in bones.

1.2 Symbols

A system of symbols was developed so that chemical reactions could be written down in shorthand form. The initial letter, or the initial plus one other letter, of an element is used to represent one atom of it, e.g. C is the symbol for carbon and Ca is the symbol for calcium. In some cases, the Latin names are used, e.g. Fe is the symbol for iron (ferrum). A full list of symbols is given on pages 209–210.

1.3 The Separation of Mixtures

The separation of mixtures is generally comparatively easy, the method chosen depending on the nature of the substances to be separated.

(a) To Obtain a Solute from a Solution

(i) *Evaporation*

A *solution* consists of a *solute* dissolved in a *solvent*, e.g. sea water is a solution of salt (the solute) in water (the solvent). If sea water is heated in the apparatus shown in Fig. 1.3 (a), the water evaporates and salt is left. The final heating may be carried out using a steam bath (Fig. 1.3 (b)) to avoid loss of salt by spitting.

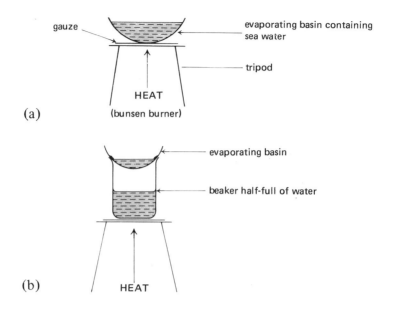

Fig. 1.3

(ii) *Crystallisation* ~~Separating Solute from solution~~

The main disadvantage of evaporating all of the solvent from a solution in order to obtain the solute is that if there are any dissolved impurities present, they will be obtained with the solid at the end of the experiment. This problem is avoided if crystallisation is carried out. The solution is evaporated to crystallisation point, i.e. the point at which crystals of solute will form on cooling the solution to room temperature. Drops of solution are removed at intervals and allowed to cool; the formation of crystals in one of these drops indicates that the solution is at crystallisation point. If the solution is allowed to cool, crystals form and can then be filtered out, washed and dried.

(b) To Separate a Solid from a Liquid

(i) *Filtration*

This method uses the apparatus shown in Fig. 1.4 to separate a solid from a liquid. The solid is left on the filter paper as the residue while the liquid passes through and is collected in the evaporating basin as the filtrate.

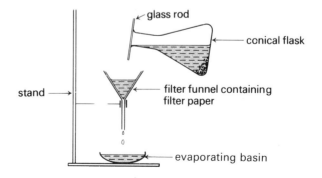

Fig. 1.4

6

A centrifuge can also be used. A tube of mixture is spun round at high speed. The solid settles to the bottom of the tube, and the liquid may then be removed by using a dropping pipette or by decantation. **Decantation** is the pouring off of a clear liquid after a substance held in suspension has settled to the bottom of the container.

(c) To Separate Two Solids

(i) *Paper Chromatography*

Paper chromatography is used to separate a mixture of similar solids dissolved in a solvent (see Example 1.3), e.g. it can be used to separate the dyes in ink. A small drop of ink is placed in the centre of a piece of filter paper (Fig. 1.5). Water passes over the ink and the dyes separate into rings of different substances. The separation depends on differences in (a) the solubilities of the dyes in the water, and (b) the tendencies of the dyes to stick to the surface of the paper. Chromatography can also be used to separate colourless substances but in this case the paper must be developed by spraying it with another chemical so that the position of the solids can be seen.

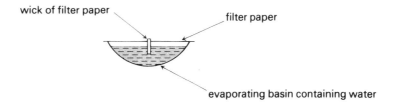

Fig. 1.5

(d) To Separate a Solvent from a Solution

(i) *Distillation*

For this separation the apparatus shown in Fig. 1.6 is used.
When the solution is boiled, the solvent changes to vapour. The vapour passes down a condenser where it is converted back to liquid and is collected as the distillate in the beaker.

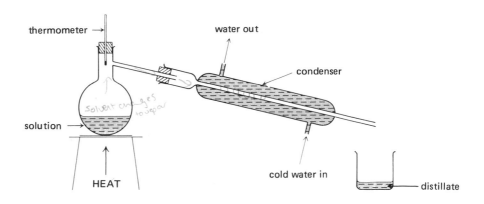

Fig. 1.6

7

(e) To Separate Two Liquids

(i) *Separating Funnel*

If two liquids are *immiscible* (i.e. they separate into two distinct layers), then the simplest way to separate them is to use a separating funnel. The lower layer may be run into a beaker and the upper layer tipped into another container.

(ii) *Fractional Distillation*

Miscible liquids (i.e. they mix together completely and do not form layers) may be separated by fractional distillation provided that their boiling points are different. When a mixture of two such liquids is heated to boiling in the apparatus shown in Fig. 1.7, the vapour of the more volatile component (i.e. the liquid with the lower boiling point) travels up the column more easily. The vapour of the component with the higher boiling point condenses more readily. Eventually the vapour escaping to the condenser consists of the more volatile component only.

Important applications of fractional distillation include:

(a) the separation of liquid air into oxygen, nitrogen, etc. (see Section 10.2);
(b) the separation of crude oil into petrol, paraffin, etc. (see Section 19.4);
(c) the manufacture of spirits (e.g. gin, whisky, brandy) (see Section 19.5).

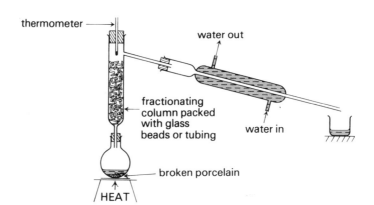

Fig. 1.7

(f) To Separate Two Gases diffusion .

Diffusion (see Section 2.3) can be used to separate two gases of differing densities.

1.4 Tests for Purity

Pure substances usually have definite sharp melting points and boiling points and the addition of only a trace of an impurity will alter these values. The presence of dissolved solid impurities raises the boiling point of a liquid and lowers the melting point of a solid, making it less sharp.

1.5   Worked Examples

Example 1.1

Which one of the following groups consists only of compounds?

A Cl_2 HCl $MgCl_2$ C_6H_5Cl
B $CaSO_4$ H_2S S_8 Na_2S
C MnO_2 PH_3 XeF_4 O_2
D $KMnO_4$ H_2SO_4 HNO_3 HCl
E NH_3 N_2 $NaNO_3$ CH_3NH_2

The answer is **D**

Cl_2, S_8, O_2 and N_2 are all the formulae for elements because each one consists of atoms of the same type.

Example 1.2

Which of the following is an element?
A brass
B emerald
C aluminium
D granite
E sand

The answer is **C**.

Brass is an alloy of copper and zinc, and sand is a compound. The other substances (**B** and **D**) are mixtures.

Example 1.3

(a) Name the chemical technique that could be used successfully to separate:
  (i) petrol from crude oil.   **(1 mark)**
Fractional distillation (see Section 1.3).

Crude oil consists of a number of different fractions including petrol. These have different boiling points and so must be separated by fractional distillation.

  (ii) the coloured components in food colouring.   **(1 mark)**
Chromatography (see Section 1.3).

(b) (i) Would the kind of experiment mentioned in (a, ii) be successful in separating ethanol from a mixture of ethanol and water?   **(1 mark)**
No

  (ii) Explain your answer.   **(1 mark)**
Ethanol and water are both colourless liquids. Paper chromatography is generally used to separate two or more coloured solids dissolved in a solvent.

(c) A pupil decides to separate powdered calcium carbonate from powdered calcium chloride by shaking the mixture with water and then filtering. Would this procedure succeed? Explain your answer.   **(2 marks)**
                                                                    (OLE)

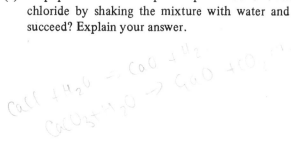

This procedure would succeed. Calcium carbonate is insoluble in water whereas calcium chloride is very soluble. On filtering the mixture, calcium carbonate would remain on the filter paper whereas calcium chloride solution would pass through.

Example 1.4

In earlier times a retort was used for carrying out distillation (see Fig. 1.8).

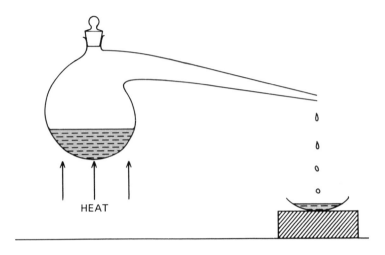

HEAT

Fig. 1.8

(i) Explain what is meant by *distillation* and state why it is an important process in chemistry. **(3 marks)**

Distillation occurs when a liquid is heated to form a vapour which is then condensed to reform the liquid. It is an important process in chemistry since it provides a means of purifying liquids.

(ii) Write down, with a few words of explanation, **two** ways in which a modern distillation set is an improvement on the above apparatus. **(2 marks)**
(OLE)

Today, distillation would be carried out using the apparatus shown in Fig. 1.6. The thermometer enables the boiling point of the distillate to be found and hence the purity of the distillate can be checked (see Section 1.4). The cold water in the condenser ensures that all the vapour changes to liquid, maximising the yield of product.

Example 1.5

Name the methods you would use to carry out each of the following separations, explaining briefly the principles upon which your methods rely:

(i) a sample of elderberry wine from a mixture of it with solid yeast residues; **(2 marks)**

Filtration (see Section 1.3). The elderberry wine will pass through the filter paper but the yeast residues, being insoluble, will remain in the filter paper.

(ii) reasonably pure ethanol from elderberry wine. **(3 marks)**

Elderberry wine will contain ethanol and water together with a number of other substances. Since these substances have different boiling points they may be separated by fractional distillation (see Section 1.3).

(iii) blue copper sulphate crystals from aqueous copper sulphate. **(2 marks)**
(OLE)

Crystallisation. Some of the water must be evaporated so that crystals will form on cooling the solution to room temperature.

Example 1.6

(i) Name two physical measurements that can be used to demonstrate that a substance is pure. **(2 marks)**

Melting point, boiling point.

(ii) Will (A) a compound and (B) a mixture behave in the same way when those measurements are made? **(2 marks)**

A compound will have a definite sharp melting point and boiling point. The melting point of a mixture will not be sharp, and its value will depend on the relative proportions of the components in the mixture. Although the boiling point of a mixture will be sharp, its value will vary.

1.6 Self-test Questions

Question 1.1

Which of the substances **A** to **E** is most likely to be an element?

A Changes colour and loses oxygen rapidly on gentle heating.
B Burns in air to form carbon dioxide and water.
C Conducts electricity as a solid and also when molten.
D Produces several fractions when distilled.
E On heating evolves a gas leaving a solid residue. (L)

Question 1.2

Use the following list of substances to answer the questions that follow: air, carbon, chlorine, hydrogen, mercury, rock salt, sodium, sulphur, water.

(a) Which metal is a liquid at room temperature and pressure? mercury
(b) Name the two elements present in common salt. sodium, chlorine
(c) Name a non-metal present in oil. carbon
(d) Which element is yellow in colour? sulphur
(e) Name an element used in street lighting. sodium
(f) Name a compound. water
(g) Name a mixture. air **(7 marks)**

Question 1.3

Complete the following statements:
(a) decantation is the process of separating a liquid from a solid sediment by pouring.
(b) centrifugation is used to separate cream from milk.
(c) A residue is the solid left on a filter paper.
(d) The filtrate is the liquid that passes through a filter paper.
(e) The process of evaporating a liquid and then condensing the vapour is known as distillation. **(7 marks)**

Question 1.4

Benzene and water are two liquids that do not mix. If placed together in a test tube, they form two separate layers. Explain how it would be possible to identify the water layer.

(2 marks)

Question 1.5

(a) Name the chemical technique which could usefully be used to separate:
  (i) the mixture of soluble dyes used to colour 'blackcurrant' sweets;
  (ii) water from a solution of sodium chloride in water. **(1 mark)**

(b) A pupil decides to separate a mixture of the solids calcium hydroxide and ammonium chloride by shaking the mixture with hot water and then filtering. Will this procedure work? Explain your answer. **(3 marks)**

(OLE)

Question 1.6

Filtration is a method of separating:
A one insoluble solid from another
B a solute from its solvent
C an insoluble solid from a liquid
D a solvent from a solution
E one soluble solid from another.

1.7 Answers to Self-test Questions

1.1 **C**.

> If substance **A** loses oxygen on heating, it must contain oxygen and something else. Substance **B** combines with oxygen from the air to give carbon dioxide and water and therefore must contain at least carbon and hydrogen. Pure substances do not produce several fractions when distilled, and therefore substance **D** must be a mixture. Substance **E** splits up on heating; elements cannot be split up into simpler substances.

1.2 (a) Mercury.
   (b) Chlorine, sodium.
   (c) Carbon or hydrogen.
   (d) Sulphur.
   (e) Sodium.
   (f) Water.
   (g) Air or rock salt.

1.3 (a) Decantation.
   (b) Centrifugation.
   (c) Residue.
   (d) Filtrate.
   (e) Evaporating, condensing, distillation.

1.4 The water layer can be identified by adding a little more water or by adding a little more benzene. Alternatively, a knowledge of the densities of the two liquids will identify the water layer since the liquid with the lower density will form the upper layer.

1.5 (a) (i) chromatography;
      (ii) distillation.
   (b) This procedure will not work. Ammonium chloride is very soluble in hot water but calcium hydroxide is slightly soluble too. Thus, although most of the calcium hydroxide would remain on the filter paper, some would dissolve and contaminate the ammonium chloride in the filtrate.

1.6 **C.**

Do not make the mistake of thinking that option **B** is correct: the solute will be dissolved in the solvent and so cannot be separated in this way.

2 Atoms and Molecules

2.1 Introduction

Matter consists of tiny particles, either atoms, or groups of atoms called molecules.

An **atom** is the smallest particle of an element that can exist and still retain the ordinary chemical properties of that element.

Many elements consist of groups of atoms joined together, e.g. oxygen, O_2, sulphur, S_8. In addition, atoms of different elements may join together in groups to form particles of a compound. These groups of atoms are called molecules.

A **molecule** is the smallest particle of an element or compound that can exist naturally and still retain the ordinary chemical properties of that element or compound.

The results of many experiments can be explained in terms of the particle theory of matter and hence they provide evidence to support it, e.g. the dissolving of a solid in a liquid or the mixing of two gases may be interpreted in terms of the intermingling of the particles of the various substances. Crystals of a particular compound all have more or less the same shape. This can easily be understood if we assume that every crystal consists of layers of particles packed together in a regular pattern.

2.2 The States of Matter

Table 2.1 summarises some differences in the structure and properties of solids, liquids and gases.

Table 2.1

	Solid	Liquid	Gas
Compressibility	Difficult to compress	Difficult to compress	Easy to compress
Molecular packing	Close	Close	Sparse
Shape	Fixed	Variable	Variable
Molecular movement	Vibrate and rotate about fixed points	Free to move	Free to move
Attractive forces	Relatively high	Relatively high	Very small
Diffusion	Very slow	Slow	Rapid

When the temperature of a solid is raised, the particles vibrate more and more violently until they are moving to such an extent that they can no longer be held in an ordered arrangement by the forces of attraction. When this happens the solid melts. Raising the temperature of a liquid increases the speed of movement of the particles until their kinetic energy is sufficient to overcome the forces of attraction between them. At this point the liquid boils. On cooling the reverse changes occur. The particles of the gas gradually slow down as the temperature falls until the forces of attraction are able to condense them together and form a liquid. Cooling the liquid causes further loss of kinetic energy until eventually the particles settle into a solid crystal where they vibrate and rotate about fixed points (see Examples 2.1, 2.2 and 2.3).

2.3 Diffusion

Gases have a tendency to spread in all directions, and this can be explained by the idea that gases are made up of moving particles (see Examples 2.1 and 2.4 and Question 2.1). The process of spreading is known as diffusion. In general, the denser the gas the slower the diffusion process.

2.4 Atomic Structure

Atoms consist of a minute nucleus, where all the positive charge and most of the mass of the atom is concentrated, surrounded by electrons. The nucleus is made up of two types of particle: protons and neutrons.

A **proton** is a positively charged particle, with mass approximately equal to that of a hydrogen atom.

A **neutron** is a neutral particle, with mass approximately equal to that of a hydrogen atom.

The **electron** is a negatively charged particle, its charge being equal but opposite to that of a proton, and its mass approximately 1/1840 of that of a proton.

Since atoms are electrically neutral, the numbers of protons and electrons in an atom must be equal.

The number of protons in the nucleus of an atom is its **atomic number** and gives its position in the periodic table.

The total number of protons and neutrons in an atom is its **mass number.**

Information about the atomic number and mass number of an element can be given with its symbol. Thus $^{12}_{6}C$ represents an atom of carbon with an atomic number of 6 and a mass number of 12.

(a) Arrangement of Electrons (see Example 2.5 and Question 2.2)

The electrons surround the nucleus and are found at varying distances from it in groups known as energy levels or 'shells'. Each shell contains electrons with similar energies and can contain up to a certain maximum number. The first shell, nearest the nucleus, can contain up to two electrons, the second up to eight and the third up to 18. For example, a potassium atom, atomic number 19, will have the electron arrangement shown in Fig. 2.1.

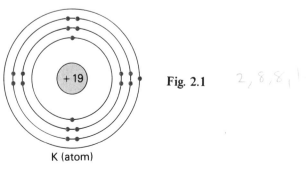

Fig. 2.1   2, 8, 8, 1

K (atom)

There is a special stability associated with having eight electrons in a shell. Thus, in a potassium atom, there are eight electrons in the third shell and one in the fourth shell even though the third is not full.

Table 3.1 (page 25) shows the atomic structures of the most common isotopes of the first 20 elements.

(b) Isotopes (see Example 2.6)

It is the number of protons in the nucleus of an atom that identifies it. However, it is possible to have atoms of the same element with different numbers of neutrons in their nuclei. They are called **isotopes**, e.g.

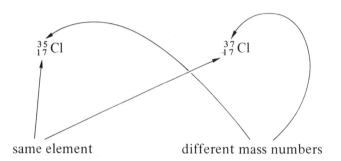

$^{35}_{17}Cl$ $^{37}_{17}Cl$

same element   different mass numbers

2.5   Radioactivity

All atoms with atomic number greater than 83, together with some of the lighter ones, are radioactive. This means that their nuclei are unstable and split up, emitting radiation which is of three distinct types — α, β and γ. When either an α- or a β-particle is emitted, the remaining nucleus usually has an excess amount of energy. This extra energy is given out in the form of γ-rays, which are like X-rays but with even shorter wavelength. When the nucleus of a radioactive atom gives off an α- or a β-particle, an atom of a new element is formed. If this new atom is stable, the disintegration process stops, but if it is radioactive, then the splitting-up process continues. Some of these radioactive nuclei (radioisotopes) occur naturally but others are made in nuclear reactors.

(a) Half-life

The **half-life** of an isotope is the time taken for the mass of radioactive material to be reduced to half its initial value (see Example 2.7).

(b) The Uses of Radioactive Isotopes

1. Nuclear power stations generate electricity from the energy produced when isotopes like uranium-235 decay. Uranium-235 absorbs neutrons to form uranium-236 and then nuclear fission occurs according to the following equation:

$$^{1}_{0}n + {}^{235}_{92}U \rightarrow [{}^{236}_{92}U] \rightarrow {}^{90}_{38}Sr + {}^{143}_{54}Xe + 3{}^{1}_{0}n$$

The neutrons emitted can cause the fission of more uranium atoms and so produce more and more neutrons. The controlled production of neutrons releases a considerable amount of heat energy which can be used to generate electricity.

2. In living animals and plants the percentage of the radioactive isotope $^{14}_{6}C$ remains constant because it is continually being replaced from the carbon dioxide in the air and from food. In dead tissue this does not happen so the percentage of $^{14}_{6}C$ gradually decreases as the isotope decays. Hence by comparing the percentage of carbon-14 in a dead sample with that in living matter the age of the sample can be found.

3. If radioactive compounds are injected into a patient the functioning of glands and the flow of blood around the body may be checked by means of a Geiger counter or some other detecting device.

4. Cancer may be cured in some cases by killing the cancerous cells by exposing them to the penetrating power of radiation.

5. γ-radiation will completely destroy bacteria and so can be used to sterilise surgical equipment.

6. Radioactive sources also have a number of industrial uses, e.g. to test for leaks in pipelines by allowing a solution of a radioactive isotope to flow through the pipe and measuring the radioactivity of the surrounding soil.

(c) The Dangers of Radiation

Radioisotopes have some potentially harmful effects. The most penetrating type is γ-radiation, but all radioactive sources must be handled with care since radiation kills all living cells. One of the problems associated with radioactive material is in the disposal of waste. Some radioisotopes have long half-lives and so can emit radiation for many years after they are no longer needed. Some radioactive waste is recycled into useful material but otherwise it must be stored in lead containers until it is no longer harmful.

2.6 Worked Examples

Example 2.1

	Solid	*Liquid*	*Gas*
Molecular packing	Close	Close	Sparse
Molecular movement	Particles vibrate and rotate about fixed points	Particles free to move	Particles free to move
Attractive forces	Relatively high	Relatively high	Very small

Use the information in the table to explain the following observations.

(a) Solids are rigid but liquids flow. **(2 marks)**

In a solid, the particles can only vibrate and rotate about fixed points whereas in a liquid the particles are free to move.

(b) Gases are easy to compress but liquids are not. **(2 marks)**

In a liquid, the particles are close together but in a gas, molecular packing is sparse. Therefore, applying pressure to a gas will push the molecules closer together but this would be difficult in a liquid.

(c) Gases fill any container in which they are placed. **(2 marks)**

In a gas, the particles are free to move, and since the attractive forces between them are small, the particles can move apart to fill the whole container.

(d) The energy required to melt 1 g of ice at 0°C is much less than that required to vaporise 1 g of water at 100°C. **(2 marks)**

In order to melt ice or vaporise water the attractive forces between the particles have to be overcome. In water the molecules are still close together but in steam they are much further apart. Therefore much more work is done in vaporising water than in melting ice.

Example 2.2

Give TWO ways in which the ions in an ionic lattice may be made free to move. **(2 marks)**
(SUJB)

Melting the ionic solid or dissolving it in water enable the ions to move.

Example 2.3

This question is about changing naphthalene from solid to liquid. Naphthalene, $C_{10}H_8$, is a white crystalline solid. 6.4 g of naphthalene crystals were weighed and placed in a test tube together with a thermometer. The tube was placed in a beaker of boiling water and a stop-clock was started. The water was kept boiling and the naphthalene crystals were stirred continuously with the thermometer. The temperature of the naphthalene was recorded every 15 seconds. The results are shown in Fig. 2.2.

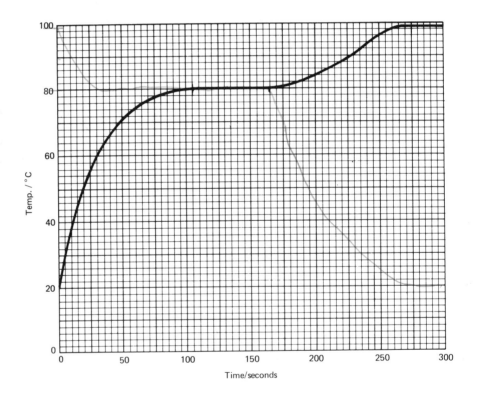

Fig. 2.2

(a) What was the boiling point of the water? **(1 mark)**

99°C.

The maximum constant temperature obtained must be the boiling point of the water.

18

(b) What was the melting point of the naphthalene? **(1 mark)**

80°C.

(c) Explain why the temperature of the naphthalene remained constant from 105 seconds to 165 seconds even though the water was kept boiling. **(2 marks)**

Since the water was kept boiling, heat was being supplied to the naphthalene. This was used up in overcoming the forces holding the particles together in the solid lattice (i.e. energy was used up in pulling the particles apart). The temperature remained constant until all of the solid had melted.

(d) Describe how the movement of the naphthalene molecules changes within the crystals during the first 30 seconds of heating. **(2 marks)**

As the temperature increases, the naphthalene molecules vibrate more vigorously about fixed points.

(e) On Fig. 2.2 sketch the curve you would expect to obtain if the test tube and naphthalene were removed from the beaker of boiling water and placed in a beaker of water at 20°C, the temperature of the naphthalene being recorded at regular intervals of time. Begin your graph at time = 0 s. **(3 marks)**

The answer is given in Fig. 2.3.

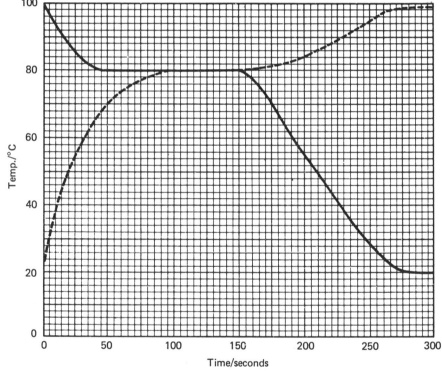

Fig. 2.3

(f) Calculate the number of moles of naphthalene molecules used in the experiment. (Relative atomic masses: C = 12, H = 1.) **(2 marks)**

(L)

See Section 5.1

Molar mass of naphthalene ($C_{10}H_8$) = (10 × 12) + (8 × 1)

$$= 128 \ g \ mol^{-1}$$

Moles of naphthalene

$$= \frac{mass}{molar \ mass}$$

$$= \frac{6.4}{128}$$

$$= 0.05$$

19

Example 2.4

Tube 1 in Fig. 2.4 shows the result of an experiment to investigate the formation of ammonium chloride by the counter-diffusion of ammonia gas and hydrogen chloride. If a similar experiment were set up as in tube 2 to investigate the reaction between ammonia gas and hydrogen bromide, at which one of the five positions shown is the solid most likely to form?

M_r (HCl) = 36.5; M_r (HBr) = 81. (AEB, 1981)

The answer is **E**.

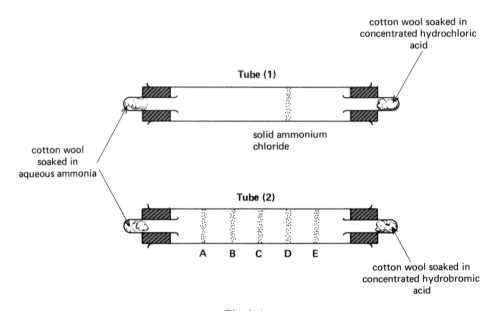

Fig. 2.4

Hydrogen bromide has a higher molar mass and a higher density than hydrogen chloride and hence it will diffuse more slowly.

Example 2.5

The table gives some information concerning the structures of four atoms, W, X, Y and Z. Work out the missing figures (a) to (l).

	Atomic number	Mass number	Number of protons	Number of neutrons	Electronic configuration
W	19	39	(a) 19	(b) 20	(c) 2,8,8,1.
X	(d) 10	20	10	(e) 10	(f) 2,8,
Y	(g) 6	(h) 12	6	6	(i) 2,8+
Z	(j) 6	(k) 14	(l) 6	8	2.4

(1 mark each)

*(a) 19   (b) 20   (c) 2.8.8.1   (d) 10   (e) 10   (f) 2.8   (g) 6   (h) 12   (i) 2.4
(j) 6   (k) 14   (l) 6*

Use the letters W, X, Y, Z when answering the following questions.

(m) Which is the atom of a noble gas? **(1 mark)**

X.

(n) Which two atoms are isotopes of the same element? **(1 mark)**

Y and Z.

(o) Which is a metal atom? **(1 mark)**

W.

> W has one electron in its outermost shell and so is in Group 1 and hence is a metal.

Example 2.6

(a) Chlorine has two common isotopes with symbols $^{35}_{17}Cl$ and $^{37}_{17}Cl$.

    (i) Using chlorine as an example, explain the meaning of the term *isotope*. **(2 marks)**

Isotopes are atoms of the same element with different numbers of neutrons in their nuclei. Thus both $^{35}_{17}Cl$ and $^{37}_{17}Cl$ have 17 protons and 17 electrons but $^{35}_{17}Cl$ has 18 neutrons whereas $^{37}_{17}Cl$ has 20 neutrons.

    (ii) Explain why chlorine, whether as the element or in its compounds, always has a relative atomic mass of about 35.5. **(2 marks)**

The relative atomic mass is the average mass of one of the atoms and has to take into account the relative abundances of the various isotopes. Natural chlorine always contains about 3/4 $^{35}_{17}Cl$ and 1/4 $^{37}_{17}Cl$.
Therefore, relative atomic mass $= (\frac{3}{4} \times 35) + (\frac{1}{4} \times 37)$
$= 35.5$

(b) Some radioactive isotopes are playing increasingly important roles in medicine. Write down the name of **one** such isotope and briefly describe its use in medicine. **(2 marks)**

(OLE)

Cobalt-60 gives off intense γ-radiation which can be used to destroy cancerous growths.

Example 2.7

(a) Explain what is meant by the term 'half-life'. **(3 marks)**

The half-life of an isotope is the time taken for the mass of radioactive material to be reduced to half its initial value.

(b) The half-life of iodine-131 is 8 days. How much iodine-131 will be left after 8 days if you have an initial mass of 1 g? **(1 mark)**

$\frac{1}{2}$ g.

(c) Name one naturally occurring radioactive metal which is used in nuclear power stations. **(1 mark)**

Uranium.

(d) Name one radioactive metal which is made in a nuclear reactor and is used in nuclear power stations. **(1 mark)**

Plutonium.

(e) Most of our electricity is obtained from nuclear power stations or by burning fuels. Give two other sources of energy which can be used to generate electricity. **(2 marks)**

Any two of moving water, wind or sunlight.

2.7 Self-test Questions

Question 2.1

(a) Why is it that in a cinema the projector beam becomes much more obvious when people in the audience begin to smoke? **(2 marks)**

(b) What do you see when smoke is lit by a beam of light and examined under a microscope? Explain your answer. **(2 marks)**

Question 2.2

The element silicon has three isotopes with mass numbers 28, 29 and 30. An accurate value for the relative atomic mass of natural silicon is 28.09.

(a) Which isotope do you consider to be most abundant? Explain your answer. **(3 marks)**

(b) For each of the isotopes, give the composition of its nucleus. **(5 marks)**

    (i) ^{28}Si

    (ii) ^{29}Si

    (iii) ^{30}Si

(c) Give the electronic structure of the isotope $^{30}_{14}$Si. **(1 mark)**

(d) From your answer to (c), would you expect silicon to form *ionic* or *covalent* compounds? Explain your answer. **(2 marks)**

(OLE)

Question 2.3

An atom contains 11 electrons, 11 protons and 12 neutrons. What is its mass number?

A 11

B 12

C 22

D 23

E 34

Use your periodic table (page 211) to answer Questions 2.4–2.6.

Question 2.4

Which atom has the electron arrangement shown in the diagram?

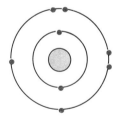

Question 2.5

How many protons are present in an atom of magnesium?

Question 2.6

How many electrons are present in a chloride ion (Cl^-)?

Question 2.7

In a radioactivity experiment the background count was found to be 14 counts per minute. At the start a radioactive specimen gave a count rate of 174 counts per minute (including background radiation). Half an hour later the count was found to be 94 counts per minute. What would be the count rate in counts per minute (including background radiation) 1 hour after the beginning of the experiment?

A 14
B 28
C 40
D 54
E 80

(L)

2.8 Answers to Self-test Questions

2.1 (a) Smoke consists of solid particles. In clean air the projector beam is invisible, but when smoke is present the beam of light collides with the particles of smoke and light is scattered in all directions, thus making the beam visible from the side.

(b) The smoke particles reflect the light, and bright specks are seen moving around continually in a random manner. The effect is due to air molecules, themselves too small to be seen, colliding with the smoke particles and causing them to bounce around in all directions.

> This sort of motion was first observed by the botanist, Robert Brown, when he looked through a microscope at pollen grains suspended in water — hence the name 'Brownian motion'.

2.2 (a) The isotope with mass number 28 will be the most abundant. The relative atomic mass is found from the average masses of the three isotopes, taking into account the relative abundances. Since the other two have masses greater than 28, they must be present to a small extent only.

(b) (i) 14 protons, 14 neutrons.
(ii) 14 protons, 15 neutrons.
(iii) 14 protons, 16 neutrons.

> Silicon has an atomic number of 14 and therefore has 14 protons. The mass number gives the number of protons + the number of neutrons and hence by subtraction the number of neutrons can be found.

(c) Silicon has 14 electrons and has the electronic structure 2.8.4.

(d) Silicon will form covalent compounds. For silicon to form ionic compounds, four electrons must be lost or gained so that silicon can have an outer shell of eight electrons. This requires too much energy and so silicon will share electrons and form covalent compounds (see Section 4.1).

2.3 **D**.

2.4 Oxygen.

2.5 12.

2.6 18

A chlorine atom has 17 electrons and so the extra electron present in the ion makes 18.

2.7 **D**.

The background count must be subtracted from each reading. Initially, the count rate due to the radioactive material will be 160 counts per minute. This drops to 80 counts per minute after half an hour. (Therefore the half-life = 1/2 hour.) After another half an hour, the actual count rate = 80/2 + 14 counts per min.

3 The Periodic Table

3.1 Introduction

The periodic table is obtained by arranging the elements in order of increasing atomic number and placing them in rows so that similar elements fall into vertical columns (see page 211). The horizontal rows are called *periods* and the vertical columns are called *groups*.

Table 3.1 shows that atoms of similar elements, such as sodium and potassium, have the same numbers of electrons in their outermost shells.

For the outer blocks of the periodic table the number of electrons in the outermost shells of the atoms is the same as the number of the group in which they are found.

Hydrogen and helium do not fit properly into any of the groups and are usually separated from the rest of the table.

When element 21 is reached the electronic structures of the atoms become more complicated and a centre block of *transition elements* has to be added.

Several of the groups have 'family' names: group I elements are called the *alkali metals*, group II elements are the *alkaline earths*, group VII elements are the *halogens* and group 0 (plus helium) the *noble gases*.

Table 3.1

Element	Atomic number	Number of neutrons	Mass number	Electronic arrangement
Hydrogen	1	0	1	1
Helium	2	2	4	2
Lithium	3	4	7	2,1
Beryllium	4	5	9	2,2
Boron	5	6	11	2,3
Carbon	6	6	12	2,4
Nitrogen	7	7	14	2,5
Oxygen	8	8	16	2,6
Fluorine	9	10	19	2,7
Neon	10	10	20	2,8
Sodium	11	12	23	2,8,1
Magnesium	12	12	24	2,8,2
Aluminium	13	14	27	2,8,3
Silicon	14	14	28	2,8,4
Phosphorus	15	16	31	2,8,5
Sulphur	16	16	32	2,8,6
Chlorine	17	18 or 20	35 or 37	2,8,7
Argon	18	22	40	2,8,8
Potassium	19	20	39	2,8,8,1
Calcium	20	20	40	2,8,8,2

3.2 The Noble Gases

Atoms of the noble gases (group 0 plus helium) have very stable electronic structures. For this reason they exist singly rather than being joined in pairs and form very few compounds. The attraction between the atoms is very small, giving the noble gases very low boiling points.

3.3 Trends in Properties Going across a Period (Outer Blocks Only)

Going across a period the elements change from metals to non-metals.

The valencies (see Section 5.1) of the elements in groups I–IV are equal to the group numbers: for the elements in groups V–VII they are usually equal to 8 minus the group number (e.g. for magnesium, in group II, the valency is 2; for nitrogen, in group V, the valency is $8 - 5 = 3$). In oxides, the valencies of the elements in groups V–VII can sometimes be equal to the group numbers (see Table 3.2).

Table 3.2

Element	Sodium	Magnesium	Aluminium	Silicon	Phosphorus	Sulphur	Chlorine	Argon
Appearance	Silvery metal	Silvery metal	Silvery metal	Black solid	Yellow solid	Yellow solid	Greenish gas	Colourless gas
Electronic structure	2,8,1	2,8,2	2,8,3	2,8,4	2,8,5	2,8,6	2,8,7	2,8,8
Most common valency	1	2	3	4	3	2	1	0
Oxide	Na_2O	MgO	Al_2O_3	SiO_2	P_4O_{10}	SO_3	Cl_2O_7	–
Melting point of oxide, K (°C)	1466 (1193)	3348 (3075)	2318 (2045)	2001 (1728)	836 (563)	303 (30)	182 (−91)	–
Bonding and structure of oxide	← Giant ionic lattice →			Giant atomic lattice (covalent)	← Covalent molecules →			–
Nature of oxide	Decreasingly basic →		Amphoteric	Increasingly acidic →				–

3.4 Trends in Properties Going Down a Group

The changes found on going down a group are less marked than those seen in going across a period, since all members of a particular group have the same number of electrons in the outermost shells of their atoms. There are three important trends:

(1) The metallic nature of the elements increases as a group is descended. This is most noticeable in group IV, which starts with the non-metal carbon and ends with the metal lead.

(2) The reactivity of *metals* increases down a group (see Example 3.2).

(3) The reactivity of *non-metals* decreases down a group (see Example 3.2).

3.5 The Transition Elements

These are all metals but, except for the first and last members of each row, they differ from the outer block metals in a number of important ways:

(1) They have relatively high melting points and densities.
(2) They form coloured compounds (e.g. copper(II) sulphate).
(3) They can have more than one valency (e.g. iron(II) and iron(III)).
(4) Both the metals and their compounds can act as catalysts (e.g. platinum in the oxidation of ammonia and manganese(IV) oxide in the decomposition of hydrogen peroxide).

3.6 Worked Examples

Example 3.1

Consider the following information about an imaginary new element named bodium, symbol Bo, which has recently been discovered.

Bodium is a solid at room temperature but is easily cut with a knife to reveal a shiny surface which rapidly tarnishes. It reacts vigorously with water liberating a flammable gas and forming a solution with a high pH value. When bodium reacts with chlorine, it forms a white solid containing 29.5% by mass of chlorine.
A_r(Bo) = 85.

(a) Calculate the empirical formula of bodium chloride. **(3 marks)**

$$
\begin{array}{ccc}
 & \text{Bo} : \text{Cl} & \\
\textit{ratio of g} & 70.5 : 29.5 & \\
\textit{ratio of mol} & \dfrac{70.5}{85} : \dfrac{29.5}{35.5} & \\
 & = 0.829 : 0.831 & \\
 & = \quad 1 : 1 &
\end{array}
$$

∴ *empirical formula of bodium chloride is* BoCl.

Empirical formula calculations are explained in Example 5.4.

(b) To which group in the Periodic Table should bodium be assigned? **(1 mark)**
Bodium should be assigned to Group I.

(c) What type of bonding is likely to be present in bodium chloride? **(1 mark)**
Ionic bonding will be present in bodium chloride.

(d) If concentrated aqueous bodium chloride were electrolysed, what would be the main products discharged at carbon electrodes? Write equations for the reactions that take place. **(4 marks)**
At the anode the product would be chlorine.

$2Cl^- (aq) - 2e^- \rightarrow Cl_2(g)$

At the cathode the product would be hydrogen.

$2H^+(aq) + 2e^- \rightarrow H_2(g)$

(e) Write an equation and name the products for the reaction between bodium and water. **(2 marks)**

$2Bo(s) + 2H_2O(l) \rightarrow 2BoOH(aq) + H_2(g)$
The products are bodium hydroxide solution and hydrogen.

(f) Write the formula for (i) bodium nitrate and (ii) bodium carbonate. For each of these compounds, state whether it would be expected to decompose at bunsen burner temperature. Name any product(s) and write an equation for any reaction which occurs.

(5 marks)

(AEB, 1983)

(i) $BoNO_3$. *Bodium nitrate would decompose at bunsen burner temperature into bodium nitrite and oxygen.*

$$2BoNO_3(l) \rightarrow 2BoNO_2(l) + O_2(g)$$

(ii) Bo_2CO_3. *Bodium carbonate would not decompose at bunsen burner temperatures.*

From the information given, bodium is similar to potassium and sodium. Answers (b)–(f) are obtained by considering what would happen in corresponding circumstances to potassium, sodium and their compounds. The two metals are in Group I of the periodic table, their chlorides are ionic (Section 4.1) and electrolysis of their aqueous chlorides gives chlorine and hydrogen (Section 7.2). Their reactions with water and the effect of heat on their nitrates and carbonates are given in Sections 13.1 and 13.2.

Example 3.2

(a) Magnesium, calcium and strontium are in Group II of the Periodic Table and have atomic numbers 12, 20 and 38 respectively.

(i) Write down the electronic configuration of magnesium and calcium. Deduce the number of electrons in the outer shell (energy level) of a strontium atom.

Mg *2, 8, 2;* Ca *2, 8, 8, 2. As strontium is also in Group II, it, too, must have two electrons in the outer shells of its atoms.*

Electronic configurations are explained in Section 2.4.

(ii) Write an equation for the reaction that would be expected to occur between strontium (symbol Sr) and water. Name the products of the reaction, and state how the reactivity of strontium would compare with that of calcium.

$$Sr(s) + 2H_2O(l) \rightarrow Sr(OH)_2(aq) + H_2(g)$$

Products are strontium hydroxide solution and hydrogen.
Strontium would be more reactive than calcium.

(iii) Explain what would be observed if aqueous sodium carbonate was added to aqueous strontium chloride. Write an equation for the reaction. **(10 marks)**

On adding aqueous sodium carbonate to aqueous strontium chloride a white precipitate of strontium carbonate would be formed.

$$SrCl_2(aq) + Na_2CO_3(aq) \rightarrow SrCO_3(s) + 2NaCl(aq)$$

Strontium will resemble calcium in its reactions and compounds (see Section 13.3). Metal atoms react by losing their outermost electrons and becoming positive ions (Section 4.1). Since atoms become larger on going down any group, the outermost electrons are at increasing distances from the nucleus and therefore less strongly attracted to it. It follows that, as the group is descended, the metal atoms more readily lose electrons to form positive ions (i.e. become more metallic) and their reactivity increases.

(b) Chlorine, bromine and iodine are three members of Group VII in the Periodic Table.
    (i) How many electrons are there in the outer shell (energy level) of a bromine atom?
Since bromine is in Group VII it must have 7 electrons in the outer shell of its atoms.

    (ii) What type of bonding would be present in the compound formed between calcium and bromine? Explain briefly the changes in electronic structure that would occur if these two elements were to combine chemically.

Ionic bonding would be present. Each calcium atom would lose its two outermost electrons and become a Ca^{2+} ion with the same electronic configuration as the nearest noble gas in the periodic table (argon, 2, 8, 8). Bromine atoms would each gain one electron to become Br^- ions with a stable octet of electrons in their outermost shells. Two Br^- ions would be needed for each Ca^{2+} ion.

    (iii) Name a halogen that would react more vigorously than bromine. **(9 marks)**
Chlorine would be more reactive than bromine.

Ionic bonding is dealt with in Section 4.1. Non-metal atoms react by gaining electrons and becoming negative ions. Since the outermost shell of electrons is further from the nucleus on going down a group, extra electrons are less strongly held. Therefore as the group is descended the non-metallic elements less readily form negative ions and become *less* reactive (compare with metals in (a)).

(c) Silicon is a non-metal and its atomic number is 14.
    (i) Draw a diagram to show the arrangement of **outer shell** electrons in a molecule of silicon tetrachloride, $SiCl_4$. (Use a small cross x to represent an electron of a silicon atom and a small circle ○ to represent an electron of a chlorine atom.)

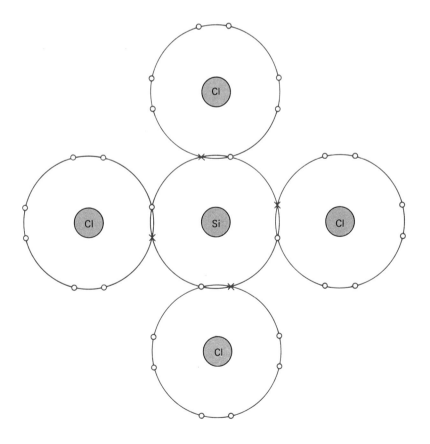

Fig. 3.1

(ii) Name the type of bonding present in the molecule, silicon tetrachloride, and state **one** physical property usually associated with compounds bonded in this way.

(5 marks)

(AEB, 1981)

Covalent bonding would be present. Compounds bonded in this way do not conduct electricity, either when molten or in solution.

Silicon has 14 electrons, arranged 2, 8, 4 and forms four covalent bonds in order to complete its octet. This is explained in Section 4.1 and Question 2.2.

Example 3.3

Three elements, X, Y and Z, are in the same period of the periodic table. The accompanying table gives some data concerning the elements and their oxides. One of the elements forms another oxide in addition to that listed.

Element	X	Y	Z
Appearance	Shiny black solid	Silvery solid	Yellow crystals
Oxide	White solid, XO_2	White solid, YO	White solid, ZO_3

Use the letters, X, Y and Z in answering the following questions. Do not try to identify the elements.

(a) What is the valency of each element in its oxide? **(3 marks)**

Valencies are: X = 4, Y = 2, Z = 6

(b) Write the letters X, Y and Z in the order in which the elements appear in the period.

(1 mark)

Y, X, Z.

Because valency = group number (Section 3.3).

(c) To which groups do the elements belong? **(3 marks)**

Y is in group II, X is in group IV, Z is in group VI.

(d) Write the formulae of the compounds which the elements would form with hydrogen.

(3 marks)

YH_2, XH_4, ZH_2.

(e) Which element would be the best conductor of electricity? **(1 mark)**

Y.

Because it is in group II and therefore is a metal.

(f) Which oxide is most likely to be ionic? **(1 mark)**

The oxide of Y.

Because Y is a metal.

Example 3.4

The Group in the Periodic Table which contains the element with eleven protons in its nucleus is

A 1

B 2

C 3

Because the atoms have eleven electrons, arranged 2, 8, 1 and the number of electrons in the outermost shell gives the group number for the outer block elements.

3.7 Self-test Questions

Question 3.1

A section of the Periodic Table is shown below for elements with atomic numbers from 3 to 18. Some of the elements are represented by letters but the letters are **not** the usual symbols for the elements.

A 3	4	5	6	7	D 8	E 9	G 10
J 11	L 12	13	M 14	15	R 16	17	18

(a) Use the letter given in the table to identify the element which:
    (i) is a noble gas   (1 mark)
    (ii) is a halogen   (1 mark)
    (iii) would react most readily with chlorine.   (1 mark)
(b) Write down the electronic configuration of:
    (i) element D   (1 mark)
    (ii) element J.   (1 mark)
What would be the formula of the compound formed if D and J were to react, and what type of bonding would be present?   (2 marks)
(c) Draw a diagram to show the electron arrangement in a molecule of the hydride of element R. (OUTER SHELLS only need be shown. Use a small cross (**x**) to represent an electron of an atom of R and a small circle (o) to represent an electron of a hydrogen atom.)   (3 marks)
                                                                     (AEB, 1982)

Question 3.2

An element whose atoms each contain 17 protons:
A is a non-metal
B has a valency of 2
C is in Group V of the Periodic Table.
D has an ionic oxide.

Question 3.3

Zirconium is a transition metal. It is therefore likely to:
A react with cold water
B form a coloured chloride
C float on water
D have only one valency

31

Question 3.4

Are the following statements true or false?
 (i) Sodium and calcium are in the same group of the Periodic Table.
(ii) Sodium and calcium atoms each have two electrons in their outer shells.

3.8 Answers to Self-test Questions

3.1 (a) (i) G (ii) E (iii) J.

Because the most reactive metals are found on the *left* of the periodic table and at the *bottom* of their groups (see Section 3.3 and Example 3.2).

 (b) (i) 2, 6 (ii) 2, 8, 1. Formula J_2D, ionic bonding (metal + non-metal – see Section 4.1).
 (c) Fig. 3.2.

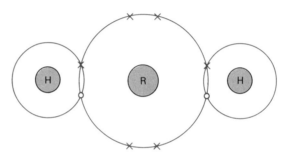

Fig. 3.2

3.2 **A.**

The element has 17 electrons, arranged 2, 8, 7, and therefore is in group VII, with valency 1. It is chlorine.

3.3 **B.**

See Section 3.5.

3.4 (i) False   (ii) False

See Section 3.1.

4 Bonding and Structure

4.1   Introduction

Bonds are formed between atoms when electrons are redistributed among those atoms to give each one the stable electronic configuration of the nearest noble gas in the periodic table. Usually this results in the formation of an octet of electrons.

An **ionic (electrovalent) bond** is formed when one or more electrons is transferred from an atom of an element on the left-hand side of the periodic table (i.e. a metal) to an atom of one on the right-hand side (i.e. a non-metal). Ionic compounds are made up of ions (see Example 4.1).

An **ion** is an electrically charged particle formed from an atom or group of atoms by the loss or gain of one or more electrons.

The number of electrons lost or gained by an atom in forming an ion is its valency. For metals outside the centre block of the periodic table, the number of electrons lost per atom is equal to the group number; for non-metals, the number of electrons gained by each atom is equal to (8 minus group number).

A **covalent bond** is formed when a pair of electrons is shared between two atoms which both need to gain electrons (i.e. non-metals).

Covalent compounds are made up of molecules (see Example 4.1).

The number of electron pairs shared by an atom is often called its valency.

4.2   Crystal Structure

The ways in which the particles are arranged in a solid can be determined by X-ray diffraction. A crystal consists of particles arranged in a repeating pattern extending in three dimensions. Such three-dimensional arrangements are called lattices.

(a) Giant Ionic Lattices

The sodium chloride lattice is shown in Fig. 4.1.

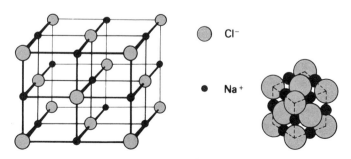

Cl⁻

Na⁺

Fig. 4.1

This arrangement extends throughout the crystal, involving millions of ions. Each cube is joined to its neighbours and the whole crystal is thus a continuous giant structure of ions or a giant ionic lattice.

(b) Giant Metallic Lattices

Metals consist of an array of positively charged ions embedded in a 'sea' of electrons. It is the forces of attraction between the positive ions and these electrons which hold the lattice together.

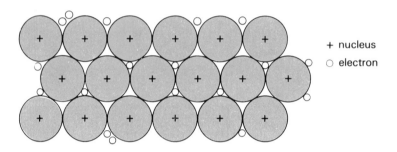

+ nucleus
○ electron

Fig. 4.2

(c) Molecular Lattices

Most covalent compounds are gases, liquids or solids which crystallise with a molecular lattice. The particles are molecules, linked by very weak attractive forces. It must be emphasised that the individual atoms *within* the molecule are joined together by strong covalent bonds. For example, in solid carbon dioxide, there are only weak forces acting between separate molecules but each carbon dioxide molecule contains a carbon atom linked by strong covalent bonds to two oxygen atoms.

(d) Giant Atomic Lattices (Sometimes Called Giant Molecular Lattices)

There are a number of substances such as silicon(IV) oxide, diamond and graphite, which have very high melting points. They consist of giant atomic lattices with strong covalent bonds acting throughout the whole crystal. The particles in these cases are atoms so they must be held together by covalent bonds, but there can be no separate molecules or the melting points would be much lower (see Example 14.1).

The properties shown by the main structural types are summarised in Table 4.1 (see Examples 4.1, 4.4 and 4.5).

Table 4.1

Property	Giant structures			Molecular structures
	Metallic	*Ionic*	*Atomic*	
m.p. & b.p.	←————————— High —————————→			Low
Solubility in organic solvents	←———— Generally insoluble ————→			Generally soluble
Solubility in water	Insoluble	Generally soluble	Insoluble	Generally insoluble
Conductors of electricity (molten)	←——— Yes ———→		←——— No ———→	
Conductors of electricity (solid)	Yes	←——————— No ———————→		
Conduction of heat	Good	←——————— Bad ———————→		

(e) Macromolecules

Some substances such as plastics, fibres and proteins are made up of molecules containing thousands of atoms. They do not have giant structures but their molecules are much bigger than the simple ones dealt with earlier. They are called macromolecules and are discussed in Chapters 15 and 19.

4.3 Oxidation and Reduction

Oxidation is:

1. the addition of oxygen to a substance;
2. the removal of hydrogen from a substance;
3. any change in which there is a loss of one or more electrons.

Reduction is the opposite of oxidation (see Example 4.7). It is:

1. the removal of oxygen from a substance;
2. the addition of hydrogen to a substance;
3. any change in which there is a gain of one or more electrons.

Electrons cannot simply disappear, and hence if one particle loses electrons (i.e. is oxidised), another must gain them (i.e. be reduced). In other words, oxidation cannot occur without reduction. Reactions involving changes of this type are often known as **redox** (*red*uction–*ox*idation) reactions.

Common oxidising agents include oxygen, chlorine, nitric acid, acidified potassium manganate(VII) and acidified potassium dichromate(VI). Common reducing agents include hydrogen, carbon and carbon monoxide.

4.4 Worked Examples

Example 4.1

(a) Name an ionic and a covalent compound and explain how each structure is formed from its elements. **(6 marks)**

(b) Show how each type of bonding influences the properties of the compound in terms of physical appearance, melting point, solubility in water and electrical conductivity.

(9 marks)

(SUJB)

(a) *Sodium chloride, NaCl, is an ionic compound. Each sodium atom (electron configuration 2,8,1) has one electron in its outermost shell while each chlorine atom (2,8,7) has seven. If the outer electron is transferred from an atom of sodium to one of chlorine, then both will have a stable octet of electrons in their outermost shells (see Fig. 4.3).*

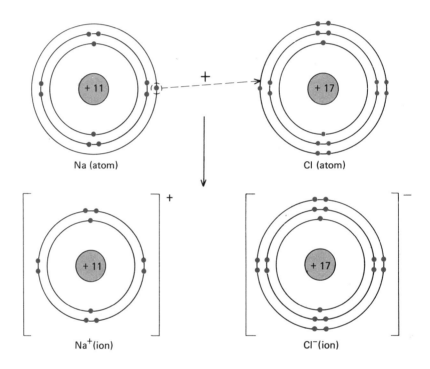

Na (atom)   Cl (atom)

Na⁺(ion)   Cl⁻(ion)

Fig. 4.3

When the sodium atom has lost its electron, it is left with a net positive charge of 1 unit since it has 11 protons (i.e. 11 positive charges) but now has only 10 electrons (i.e. 10 negative charges). Similarly, the chlorine atom acquires unit negative charge since it contains the original 17 protons but now has 18 electrons. These charged particles, called ions, strongly attract one another, the Na⁺ and Cl⁻ ions being arranged in a repeating pattern in three dimensions (see Section 4.2).

Methane, CH₄, is an example of a covalent compound. Carbon (electron configuration 2,4) has 4 electrons in its outermost shell and would have to form the C⁴⁺ ion in order to attain the stable electronic configuration of helium. So much energy is required to form the C⁴⁺ ion that the carbon and hydrogen atoms combine by an easier method: they share pairs of electrons. Each atom contributes one electron to a shared pair and in this way they all achieve stable octets (Fig. 4.4).

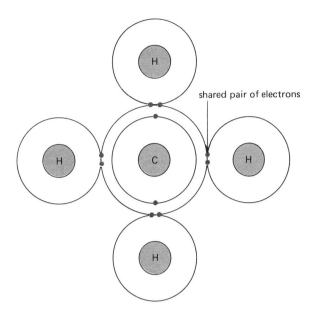

shared pair of electrons

Fig. 4.4

The atoms are held together by the attraction of the two nuclei for the shared pair of electrons. The particles formed are molecules which are free to move in all directions.

(b) *Sodium chloride is a solid with a high melting point because a large amount of energy has to be provided to overcome the strong forces of electrostatic attraction holding the oppositely charged ions together. On the other hand, the forces of attraction between neighbouring methane molecules are weak and so the melting point is low — methane is a gas at room temperature.*

Before we can consider the solubility of a compound in water, we need to know something about the water molecule. Oxygen is very electronegative. This means that an oxygen atom attracts electrons and tends to pull them towards itself in any molecule in which it is found. Thus the electron pairs making up the covalent bonds between the oxygen atom and the two hydrogen atoms in a water molecule are not shared equally but are displaced towards the oxygen. The molecule is polar, with a small negative charge on the oxygen atom and small positive charges on the hydrogen atoms.

The attraction between the ions and the water molecules is strong enough to overcome the forces holding the ions together in the crystal, so sodium chloride dissolves and the ions are carried off into solution surrounded by a 'jacket' of water molecules.

Methane is insoluble in water because the attraction of the water molecules for one another is far greater than their attraction for non-polar methane molecules.

Organic solvents, e.g. methylbenzene, do not generally dissolve ionic solids. The molecules of these solvents are non-polar or only slightly polar and hence the forces of attraction between them and the ions in the crystal are weak. For a covalent solid, the forces of attraction between the individual molecules in the lattice are comparable with those between the solvent molecules. The lattice is easily penetrated by the solvent molecules, and a solution is formed. You need to remember the phrase 'like dissolves like'. A highly polar solvent such as water dissolves polar (ionic) solids, whereas a non-polar solvent such as methylbenzene dissolves non-polar (covalent) solids.

Sodium chloride does not conduct an electric current when solid because the ions are only able to vibrate about fixed positions in the crystal lattice and cannot migrate towards the electrodes. When sodium chloride is melted or dissolved in water, the ions are free to move and hence carry an electric current. Covalent compounds cannot conduct an electric current because they do not contain ions.

Example 4.2

Element X has an atomic number of 12. The formula of its chloride will be

A X_2Cl
B XCl
C XCl_2
D XCl_3
E XCl_4

The answer is C.   (AEB, 1982)

The electron configuration of X must be 2,8,2. This means that X must lose two electrons in order to attain a stable octet, but each atom of chlorine (2,8,7) can only accept one. Hence the X atom must give its two electrons to two chlorine atoms as illustrated in Fig. 4.5.

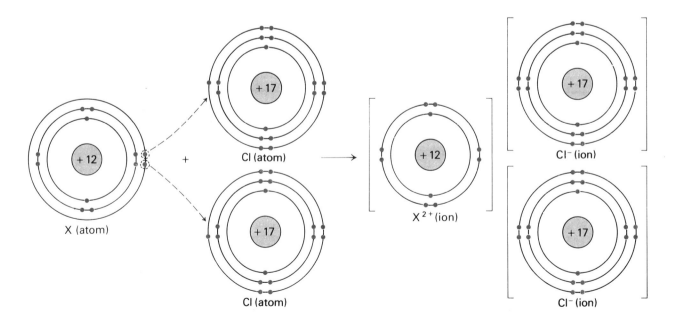

Fig. 4.5

Example 4.3

Which one of the following pairs of ions are present in silver chloride?

A Ag^+ and Cl^-
B Ag^{2+} and Cl^{2-}
C Ag^+ and Cl^{2-}
D Ag^- and Cl^+
E Ag^{2-} and Cl^{2+}

The answer is **A.**

Example 4.4

(a) Name two types of particle, other than electrons, which can be found in an atom.

(2 marks)

Protons and neutrons.

(b) Where are these particles usually found in the atom? (1 mark)

Nucleus.

(c) Calcium has an atomic number of 20 and a mass number of 40. State how many of each type of particle are present in an atom of calcium. (3 marks)

Protons   20
Neutrons   20
Electrons   20

(d) By means of a diagram (Fig. 4.6) show how the electrons are arranged in a calcium atom.

(2 marks)

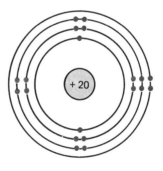

Fig. 4.6

(e) Calcium combines with oxygen (atomic number = 8) to form calcium oxide. How are the electrons arranged in an oxygen atom? (2 marks)

There are 2 electrons in the first shell and 6 electrons in the second shell.

(f) How many electrons does calcium need to lose in order to possess a stable octet in its outer shell? (1 mark)

2.

(g) How many electrons does oxygen need to gain in order to possess a stable octet in its outer shell? (1 mark)

2.

(h) By means of a diagram (Fig. 4.7) show how calcium oxide is formed from its elements.

(2 marks)

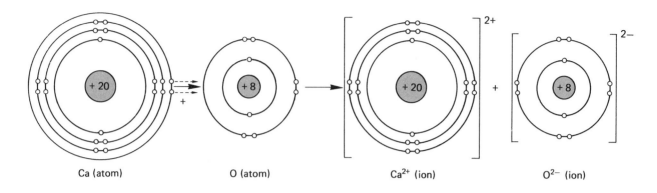

Fig. 4.7

(i) Explain, either in words or by means of a diagram (Fig. 4.8), how the electrons are arranged in a molecule of water. **(2 marks)**

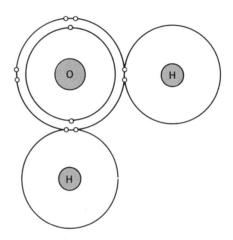

Fig. 4.8

Each atom donates one electron to a shared pair of electrons, and in this way the oxygen atom gains a stable octet of electrons and each hydrogen atom has a full outer shell of two electrons.

(j) Give two properties which are characteristic of ionic compounds. **(2 marks)**
High melting point.
Conduct an electric current when molten.

(k) How do the properties of covalent compounds differ from this? **(2 marks)**
Covalent compounds generally have low melting points and will not conduct an electric current.

Example 4.5

The table opposite gives some data for elements *A* to *E*. These letters are not the chemical symbols for the elements.

Element	Melting point (°C)	Boiling point (°C)	Heat of fusion (kJ mol^{-1})	Heat of vaporisation (kJ mol^{-1})
A	− 219	− 183	0.22	3.4
B	650	1117	8.95	128.7
C	− 7	58	5.29	15.0
D	232	2687	7.20	290.4
E	1540	2887	15.36	351.0

(a) Which element is a gas at room temperature? **(1 mark)**

A.

(b) Which element is a liquid at room temperature? **(1 mark)**

C.

(c) Will element *E* be a solid, a liquid or a gas at 1000°C? **(1 mark)**

Solid.

(d) Which of these elements do you think are composed of simple molecules? **(1 mark)**

A and C.

(e) Give TWO of the elements which could be metals. **(1 mark)**

Any two of B, D and E.

(f) Pure copper will not withstand a great bending force. What modification of the grain structure of copper will increase its strength? **(1 mark)**

Copper is flexible because the layers of ions can slide past one another without greatly altering the forces of attraction between them and the mobile 'sea' of electrons. Areas of close packing (i.e. where there are perfect rows of atoms) are called grains. Copper will not bend easily if the grains are small. Grains can be broken up by cooling the molten metal suddenly or by introducing a few atoms of a different size.

(g) Explain why copper is a good conductor of electricity. Suggest two reasons for using aluminium instead of copper for high-voltage overhead power cables. **(3 marks)**

The 'sea' of electrons normally moves in the lattice with random motion but if a potential difference is applied, it will move easily through the metal in one direction, this flow of electrons being an electric current.

Aluminium is much cheaper than copper. It is also of lower density.

Example 4.6

Each of the sets of properties (a)-(c) refers to a different substance. Complete the table by giving the name of **one** substance whose properties fit the description given and state the type of bonding present.

Structure of substance	Name	Type of bond
(a) Consists of molecules containing only two atoms which are <u>different</u> from each other	(i)	(ii) **(2 marks)**
(b) A giant molecule. The attractive forces are between atoms within the giant molecule and are symmetrical	(iii)	(iv) **(2 marks)**
(c) The forces of attraction are between positive ions and electrons	(v)	(vi) **(2 marks)**

(AEB, 1981)

(i) *Hydrogen chloride.*   (ii) *Covalent.*
(iii) *Diamond.*   (iv) *Covalent.*
(v) *Magnesium.*   (vi) *Metallic.*

Example 4.7

(a) Describe the processes of oxidation and reduction in terms of electron transfer.

(2 marks)

Oxidation is any change in which there is a loss of one or more electrons.
Reduction is any change in which there is a gain of one or more electrons.

(b) Give **one** equation in each case to illustrate the following conversions. State whether the process is an oxidation or a reduction.
 (i) Non-metal ions to a molecule.

Equation   **(2 marks)**
$2Br^- (aq) + Cl_2 (aq) \rightarrow Br_2 (aq) + 2Cl^- (aq)$
Type of process   **(1 mark)**
Oxidation of Br^- *ions*
 (ii) Metal ion to metal ion of higher charge.
Equation   **(2 marks)**
$2Fe^{2+}(aq) + Cl_2(aq) \rightarrow 2Fe^{3+}(aq) + 2Cl^- (aq)$
Type of process   **(1 mark)**
Oxidation of Fe^{2+} *ions*   (AEB, 1981)

Example 4.8

(a) Name a gas that will reduce copper(II) oxide.   **(1 mark)**
Hydrogen.

(b) Describe what you see when the reaction occurs.   **(2 marks)**
Black copper(II) oxide changes to reddish brown copper on heating.

(c) Write an equation for the reaction.   **(1 mark)**
$CuO(s) + H_2(g) \rightarrow Cu(s) + H_2O(g)$

(d) Will this reagent reduce magnesium oxide? Explain your answer.   **(2 marks)**
No, since hydrogen is below magnesium in the reactivity series.

4.5 Self-test Questions

Questions 4.1–4.6 concern the following elements and compounds:
A lead bromide
B neon
C sulphur
D aluminium
E diamond.

Which substance:

Question 4.1

Is composed of a giant structure of positive and negative ions?

Question 4.2

Has the lowest melting point?

Question 4.3

Will have a structure similar to that of silicon?

Question 4.4

Consists of separate atoms?

Question 4.5

Conducts electricity when molten but not when solid?

Question 4.6

Is a yellow, brittle, non-conducting solid with a fairly low melting point?

(1 mark each)

Question 4.7

Substance X has a molecular lattice and substance Y has an ionic lattice. Which one of the following statements about X and Y is probably correct?

A Y has a lower melting point than X.
B Y is more volatile than X.
C Molten Y conducts electricity but molten X does not.
D Y is insoluble in water but X is soluble.
E Y is soluble in tetrachloromethane but X is insoluble.   (AEB, 1981)

Question 4.8

Iodine and naphthalene are examples of substances that exist as molecules.
  (i) State two liquids in which they would be expected to dissolve readily.   **(2 marks)**
  (ii) Would the same solvents be expected to dissolve sodium chloride? Give a reason for your answer.   **(2 marks)**

Question 4.9

The reaction between aluminium and iron oxide is exothermic.

$$Fe_2O_3(s) + 2Al(s) \rightarrow Al_2O_3(s) + 2Fe(s)$$

  (a) What do you understand by the term 'exothermic'?   **(1 mark)**
  (b) Name the reducing agent in this reaction.   **(1 mark)**

4.6 Answers to Self-test Questions

4.1 **A.**

4.2 **B.**

4.3 **E.**

4.4 **B.**

4.5 **A.**

4.6 **C.**

4.7 **C.**

4.8 (i) Iodine and naphthalene have molecular lattices and dissolve in organic solvents, e.g. methylbenzene, petroleum spirit.
   (ii) No. Sodium chloride is ionic and so requires a polar solvent (see Example 4.1).

4.9 (a) The reaction gives out heat.
   (b) Aluminium.

5 Moles, Formulae and Equations

5.1   Introduction

The **relative atomic mass**, A_r, of an element is the number of times that the average mass of one of its atoms is greater than 1/12 of the mass of an atom of $^{12}_{6}C$*. Since it is a ratio, relative atomic mass has no units. Values of relative atomic masses are given on pages 209–210.

A **mole** of any substance is the amount of it which contains as many particles (atoms, molecules, etc.) as there are carbon atoms in 12 g of $^{12}_{6}C$*. The abbreviation for mole is 'mol'.

The number of particles per mole of any substance is called the **Avogadro Constant**, L. Its value is 6.02×10^{23} mol^{-1} (mol^{-1} means 'per mol').

The **molar mass** of a substance, M, is the mass of one mole of it. The units are g mol^{-1} (or g/mol).

For practical purposes, particularly in calculations, remember that one mole of *atoms* of any element:

(a) contains 6.02×10^{23} atoms;

(b) has a mass in grams which is numerically equal to the relative atomic mass of the element.

From (a) it follows that to convert an amount in moles to a number of atoms we must multiply by 6.02×10^{23},

i.e. **number of atoms = mol** $\times$ L – see Example 5.3.

To convert a number of atoms to an amount in mol we must divide by 6.02×10^{23},

i.e. **mol = number of atoms/**L.

From (b), conversion of a mass in grams to an amount in moles involves dividing by the molar mass,

i.e. **mol = mass/molar mass** – see Examples 5.1 and 5.3.

Conversion of an amount in moles to a mass in grams involves multiplying the amount in moles by the molar mass,

i.e. **mass = mol** $\times$ **molar mass** – see Example 5.2.

The **empirical formula** of a compound shows the simplest ratio of the numbers of atoms of the different elements in it.

A **radical** is a group of atoms which occurs in compounds but which cannot exist on its own. For example, the sulphate radical, SO_4, is found in all sulphates but there is no substance which is just called 'sulphate'.

Empirical formulae can be calculated from experimental results (see Example 5.4), or worked out using valencies (see Example 5.7). The **valency** of an ion

---

*For an explanation of the figures in $^{12}_{6}C$, see page 15.

(Section 4.1) shows its combining power, and combination between ions takes place in such proportions that the sum of the valencies of the positive ions or hydrogen atoms in a formula equals the sum of the valencies of the negative ions or non-metal atoms (see Question 5.2.(g)). For example, in aluminium nitrate the ratio of aluminium ions (valency 3) to nitrate ions (valency 1) is 1:3, making the empirical formula $Al(NO_3)_3$.

$$Al(NO_3)_3$$
valencies: 3 (3×1)

Notice that if a radical appears more than once in a formula it must be enclosed in a bracket.

Table 5.1 shows the valencies of the commonly encountered ions.

Table 5.1

Metal or other positive ion		Valency	Non-metal or other negative ion	
Ammonium	NH_4^+		Br^-	Bromide
Hydrogen	H^+		Cl^-	Chloride
Potassium	K^+	1	OH^-	Hydroxide
Silver	Ag^+		I^-	Iodide
Sodium	Na^+		NO_3^-	Nitrate
Barium	Ba^{2+}			
Calcium	Ca^{2+}			
Copper (in copper(II) compounds)	Cu^{2+}			
Iron (in iron(II) compounds)	Fe^{2+}	2	CO_3^{2-}	Carbonate
Lead (in lead(II) compounds)	Pb^{2+}		O^{2-}	Oxide
			SO_4^{2-}	Sulphate
Magnesium	Mg^{2+}			
Zinc	Zn^{2+}			
Aluminium	Al^{3+}			
Iron (in iron(III) compounds)	Fe^{3+}	3	PO_4^{3-}	Phosphate(V)

The **relative molecular mass**, M_r, of an element or compound is the number of times that the average mass of one of its molecules is greater than 1/12 of the mass of an atom of $^{12}_6C$. Like relative atomic mass, this is a ratio and has no units.

The relative molecular mass of a substance may be calculated by adding together the relative atomic masses of the individual atoms in one molecule of it, e.g. the relative molecular mass of carbon dioxide, CO_2, is $(12 + 16 + 16) = 44$.

For compounds with giant structures of atoms or ions (see Section 4.2) the term 'relative molecular mass' applies to a *formula unit* of the substance. Thus for potassium chloride, empirical formula KCl, the relative molecular mass is $(39 + 35.5) = 74.5$.

The **molecular formula** of a compound shows the actual numbers of atoms of the different elements in one molecule of it. For example, a molecule of hydrogen peroxide consists of two hydrogen atoms and two oxygen atoms and so its molecular formula is H_2O_2. (Note that its *empirical* formula is HO.) Unlike relative molecular mass, the term 'molecular formula' must be applied *only* to substances that consist of separate molecules.

The definitions of the mole and Avogadro constant given on page 45 apply to molecules and formula units, as well as to atoms. Thus, just as a mole of atoms of an element contains 6.02×10^{23} atoms and has a mass in grams which is numerically equal to its relative atomic mass, so a mole of a *compound* contains 6.02×10^{23} *molecules* (or formula units) and its mass in grams is numerically equal to its relative *molecular* mass. Conversions between masses, moles and numbers of particles follow the same paths as those given for atoms (see Examples 5.1, 5.3 and 5.6).

5.2 Chemical Equations

A chemical equation shows the relative numbers of atoms and molecules taking part in a chemical reaction. It can also show whether the various substances involved are in the solid (s), liquid (l) or gaseous (g) state, or are dissolved in water (aq). To write an equation you should:

1. *Know* what the reactants and products are.
2. Write down the formulae and states of the reactants and products, leaving space in front of each formula for balancing.
3. Work through the equation from left to right, checking that the same numbers of atoms of the various elements appear on both sides.

For example, when magnesium burns in oxygen it forms magnesium oxide. Applying step (2) we have:

$$Mg(s) + O_2(g) \rightarrow MgO(s)$$

The magnesium atoms balance but there are two oxygen atoms on the left and only one on the right. Therefore we try a 2 in front of the MgO(s).

$$Mg(s) + O_2(g) \rightarrow 2MgO(s)$$

The 2 in front of the MgO(s) doubles the number of magnesium atoms as well as the number of oxygen atoms so, in order to balance the whole equation we need a 2 in front of the Mg(s), giving:

$$2Mg(s) + O_2(g) \rightarrow 2MgO(s)$$

5.3 Ionic Equations

These include only those ions that *change* in some way during a reaction. Ions that do not change are called *spectator ions* and are omitted.

A useful working rule to remember when writing ionic equations is that *metallic compounds, ammonium compounds* and *dilute acids* are ionic, all other compounds being covalent. (This is an oversimplification but will suffice at this level.) You should also remember that the charges on the ions in *solids* are usually omitted, though they are still there.

To write an ionic equation you should:

1. Write a normal chemical equation for the reaction.
2. Write the equation in terms of ions, remembering that in aqueous solution the ions are completely separate from one another.
3. Rewrite the equation, omitting the spectator ions.

For example, when aqueous solutions of magnesium sulphate and barium chloride are mixed, a white precipitate of barium sulphate is formed and a solution of magnesium chloride remains. Applying the steps above, we have:

1. $BaCl_2(aq) + MgSO_4(aq) \rightarrow BaSO_4(s) + MgCl_2(aq)$
2. $Ba^{2+}(aq) + 2Cl^-(aq) + Mg^{2+}(aq) + SO_4^{2-}(aq) \rightarrow BaSO_4(s) + Mg^{2+}(aq) + 2Cl^-(aq)$
3. Clearly, the $Mg^{2+}(aq)$ and $Cl^-(aq)$ ions are spectator ions and take no part in the reaction. The ionic equation is therefore

$$Ba^{2+}(aq) + SO_4^{2-}(aq) \rightarrow BaSO_4(s)$$

Note For simple precipitation reactions of this kind you may find it easier simply to write first of all the formula of the precipitate on the right-hand side and then the formulae of the ions which go to make it up on the left-hand side, thus missing out steps (1) and (2) entirely (see Example 20.5(a), parts (i) and (ii)).

5.4 Calculating Reacting Masses from Equations

The steps in this type of calculation are shown in Example 5.8.

5.5 Worked Examples

Example 5.1

If 0.5 mol of a hydrated salt contains 63 g of water, how many moles of water of crystallisation are contained in 1 mol of salt?
$M_r(H_2O) = 18$.
A 1
B 2
C 3
D 7
E 10

(AEB, 1981)

The answer is **D**.

0.5 mol of hydrated salt contains 63 g H_2O
∴ 1 mol of hydrated salt contains 126 g H_2O
Molar mass of water = 18 g mol^{-1}
∴ Mol H_2O per mol of salt = 126/18 = 7 (amount in mol = mass/molar mass — see Section 5.1)

Example 5.2

How many grams of nitrogen gas, N_2, contain as many atoms as there are in two moles of oxygen gas, O_2? (N = 14).
A 70
B 56
C 42
D 35
E 28

(NISEC)

The answer is **B**.

> In 2 mol of O_2 there are 4 mol of O atoms
> Molar mass of N atoms = 14 g mol^{-1}
> Mass = mol × molar mass (see Section 5.1)
> ∴ 4 mol of N atoms has mass 4 × 14 = 56 g.

Example 5.3

(a) The Avogadro constant, L, is interpreted as the number of specified particles in 1 mol of a substance. State, in terms of L, the number of:

  (i) atoms in 20 g of argon   (1 mark)

0.5L

> No of atoms = mol of atoms × L = 20/40 × L
> (molar mass of Ar = 40 g mol^{-1})

  (ii) molecules in 20 g of hydrogen   (1 mark)

10L

> No of molecules = mol of H_2 × L = 20/2 × L
> (molar mass of H_2 = 2 g mol^{-1})

  (iii) hydrated protons in 1 dm^3 of aqueous sulphuric acid of concentration 1 mol/dm^3.   (1 mark)

2L

> 1 mol of H_2SO_4 gives 2 mol of H$^+$(aq)

  (iv) electrons needed to discharge 1 mol of Cu^{2+} ions.   (1 mark)

2L

> Each Cu^{2+} ion needs 2 electrons to discharge it.

(b) (i) 2.3 g of a metal **T** (A_r(**T**) = 69) displaced 3.2 g of copper from excess aqueous copper(II) sulphate. Calculate the number of moles of each metal involved in the reaction and hence determine the charge on an ion of metal **T**.   (4 marks)

Mol of Cu = mass/molar mass = 3.2/64 = 1/20; mol of T = 2.3/69 = 1/30
1/20 mol Cu is displaced by 1/30 mol T
∴ 3 mol Cu are displaced by 2 mol T
3 mol Cu^{2+} ions require 6 mol electrons for discharge
∴ 2 mol T atoms supply 6 mol electrons
∴ 1 T atom must supply 3 electrons and the charge on an ion of T is 3+

  (ii) Write an ionic equation for the reaction.   (1 mark)

2T(s) + 3Cu^{2+}(aq) → 3Cu(s) + 2T^{3+}(aq)

(AEB, 1983)

Example 5.4

2.38 g of tin was converted to 3.02 g of tin oxide. What is the formula of the oxide?
$A_r(Sn) = 119; A_r(O) = 16$.

A Sn_4O

B Sn_2O

C SnO

D SnO_2

E SnO_4 (AEB, 1983)

The answer is **D**.

Mass of oxygen combined with 2.38 g of tin = (3.02 − 2.38) = 0.64 g

$$
\begin{array}{lll}
 & Sn & : O \\
\text{ratio of g} & 2.38 & : 0.64 \\
\text{ratio of mol} & \dfrac{2.38}{119} & : \dfrac{0.64}{16} \quad \text{(No of mol = mass/molar mass)} \\
 & = 0.02 & : 0.04 \quad \text{(Divide by smallest to convert to whole numbers)} \\
 & = 1 & : 2
\end{array}
$$

∴ Empirical formula is SnO_2

Example 5.5

(a) A hydrocarbon contains 82.8% carbon. Determine the empirical (i.e. simplest) formula for this compound.   **(5 marks)**

A hydrocarbon is made up of hydrogen and carbon only, so if the percentage of carbon is 82.8, the rest (17.2%) must be hydrogen.

$$
\begin{array}{lll}
 & C & : H \\
\textit{ratio of mol} & \dfrac{82.8}{12} & : \dfrac{17.2}{1} \\
 & = 6.90 & : 17.2 \\
 & = 1 & : 2.49 \approx 2 : 5. \; \therefore \textit{ Empirical formula is } C_2H_5.
\end{array}
$$

(b) A rough estimate puts the relative molecular mass of this hydrocarbon between 50 and 60. What is the actual value? Explain your answer.   **(3 marks)**
                                                                           (OLE)

Relative mass of $C_2H_5 = (24 + 5) = 29$
Molecular formula must be a whole number × C_2H_5, *i.e.* C_2H_5, C_4H_{10}, *etc.*
∴ M_r *must be a whole number* × 29. *In this case it must be 58.*

Example 5.6

On heating 8.0 g of hydrated copper(II) sulphate, $CuSO_4.xH_2O$, 5.1 g of anhydrous salt remained. The formula of the hydrated salt was

A $CuSO_4.2H_2O$

B $CuSO_4.3H_2O$

C $CuSO_4.4H_2O$

D $CuSO_4.5H_2O$

E $CuSO_4.7H_2O$.

The answer is **D**.

The mass of water of crystallisation evolved = $(8.0 - 5.1) = 2.9$ g
Molar mass of anhydrous $CuSO_4 = (63.5 + 32 + 64) = 159.5$ g mol^{-1}

Molar mass of water $= (16 + 2) = 18$ g mol^{-1}

$$
\begin{array}{ccc}
& CuSO_4 & : & H_2O \\
\text{ratio of mol} & \dfrac{5.1}{159.5} & : & \dfrac{2.9}{18} \\
= & 0.032 & : & 0.16 \\
= & 1 & : & 5
\end{array}
$$

$\therefore$ Empirical formula is $CuSO_4.5H_2O$

Example 5.7

Metal M, with atomic number 3, has an oxide with formula:
A M_3O_2
B M_2O
C M_2O_3
D MO_2
E MO_3

The answer is **B.**

M has the electronic configuration 2,1 and therefore is in group I of the periodic table with valency 1. Oxygen has valency 2.

Example 5.8

2.8 g of iron reacted completely with chlorine according to the equation

$$2Fe(s) + 3Cl_2(g) \rightarrow 2FeCl_3(s) \qquad (Fe = 56; FeCl_3 = 162.5)$$

The mass of iron(III) chloride formed would be about:
A 2.8 g
B 4.0 g
C 5.6 g
D 8.1 g
E 16.2 g   (L)

The answer is **D.**

$$2Fe(s) + 3Cl_2(g) \rightarrow 2FeCl_3(s)$$

$$
\begin{array}{lll}
\text{Reacting masses} & 2 \times 56 \text{ g} & 2 \times 162.5 \text{ g} \\
\therefore & 112 \text{ g Fe form} & 325 \text{ g FeCl}_3 \\
\therefore & 2.8 \text{ g Fe form} & \dfrac{325}{112} \times 2.8 \text{ g FeCl}_3 \\
& & \approx 8.1 \text{ g FeCl}_3
\end{array}
$$

5.6 Self-test Questions

Question 5.1

6.02×10^{23} is the number of:
A molecules in 1.0 g of hydrogen gas
B ions in 58.5 g of sodium chloride
C molecules in 16 g of methane
D molecules in 7 g of nitrogen

Question 5.2

Some iodine was added to a weighed sample of powdered metal, M, in a test tube and the combined mass was found. Ethanol was added and the metal and iodine were seen to react. When the reaction had finished, all of the iodine had gone but some metal was left in the bottom of the tube. The ethanol was removed and the metal washed with fresh ethanol. It was then dried and weighed.

The results were

Mass of test tube + M at beginning   = 14.77 g
Mass of test tube + M + iodine   = 15.27 g
Mass of test tube + metal at end   = 14.64 g
$A_r(M) = 65$; $A_r(I) = 127$.

(a) What happened to the compound formed by the reaction?   (1 mark)
(b) What mass of M reacted with the iodine?   (1 mark)
(c) What fraction of a mole of M atoms is this?   (2 marks)
(d) What mass of iodine reacted?   (1 mark)
(e) What fraction of a mole of iodine atoms is this?   (2 marks)
(f) Write down the empirical formula of the compound formed between M and iodine.
   (2 marks)
(g) What is the valency of M?   (1 mark)

Question 5.3

Are the following statements true or false?
(i) Magnesium oxide has the formula MgO.
(ii) Magnesium oxide is formed by the combination of 1 mol of magnesium atoms with 1 mol of oxygen molecules.

Question 5.4

A compound of relative molecular mass 180 and empirical formula CH_2O has the molecular formula:
A CH_2O
B $C_2H_4O_2$
C $C_3H_6O_3$
D $C_4H_8O_4$
E $C_6H_{12}O_6$
$(A_r(H) = 1; A_r(O) = 16; A_r(C) = 12)$

52

Question 5.5

The percentage by mass of water of crystallisation in hydrated sodium sulphate, $Na_2SO_4.10H_2O$, is about

A 10

B 20

C 28

D 56

E 70

Question 5.6

A metal M forms a chloride MCl_2 of relative formula mass 95. It reacts with aqueous sodium hydroxide to give a precipitate of metal hydroxide. What is the relative formula mass of the hydroxide? (H = 1, O = 16 and Cl = 35.5)

A 24

B 40

C 41

D 58

E 95

(NISEC)

5.7 Answers to Self-test Questions

5.1 C.

---

1 g of hydrogen contains 1 mol of H atoms but only $\frac{1}{2}$ mol H_2
58.5 g is 1 mol of NaCl, containing 1 mol Na^+ and 1 mol Cl^-.
Methane is CH_4, so 16 g is 1 mol.
7 g of nitrogen contains $\frac{1}{4}$ mol N_2.

5.2 (a) It dissolved in the ethanol and was removed.
(b) $(14.77 - 14.64) = 0.13$ g
(c) $0.13/65 = 0.002$ mol
(d) $(15.27 - 14.77) = 0.50$ g
(e) $0.50/127 = 0.0039$ mol
(f) mol M : mol I = $0.002 : 0.0039 \approx 1 : 2$. ∴ empirical formula is MI_2.
(g) Valency of M is 2 because valency of I is 1.

5.3 (i) True (ii) False

---

1 mol of magnesium atoms combines with 1 mol of oxygen *atoms*, not O_2 molecules.

5.4 E.

Relative empirical formula mass of CH_2O is $(12 + 2 + 16) = 30$
Relative molecular mass is 180, which is 6×30
∴ Molecular formula is $6 \times$ empirical formula.

5.5 D.

$M_r (Na_2SO_4 \cdot 10H_2O) = (2 \times 23) + 32 + (4 \times 16) + (10 \times 18) = 322$

Percentage of water $= \dfrac{10 \times 18}{322} \times 100 = 55.9 \approx 56.$

5.6 D.

A_r (metal) must be $95 - (2 \times 35.5) = 24$

If the chloride is MCl_2, the hydroxide must be $M(OH)_2$

$M_r (M(OH)_2)$ is $24 + 2 \times (16 + 1) = 58.$

6 Gas Volumes

6.1 Introduction

Avogadro's principle states that equal volumes of all gases at the same temperature and pressure contain the same number of molecules.

Thus if two *volumes* of gas A combine with one *volume* of gas B (measured at the same temperature and pressure), then two *molecules* of gas A must combine with one *molecule* of gas B (see Examples 6.1 and 6.3).

The **molar volume** of a gas is the volume of one mole of it. Since one mole of *any* gas contains the same number of molecules (6.02×10^{23}), all gases must have the same molar volume, at the same temperature and pressure. At room temperature and pressure the value is usually taken to be 24 dm^3 mol^{-1}.

Just as dividing the *mass* of a substance by its molar *mass* gives the amount of the substance in moles (Section 5.1), so dividing the *volume* of a gas by its molar *volume* also gives an amount in moles (see Example 11.7(b)(iii)).

Multiplying an amount of gas in moles by its molar volume gives the volume of the gas.

The **atomicity** of an element is the number of atoms in one molecule of it, e.g. hydrogen, H_2, oxygen, O_2, and nitrogen, N_2, are **diatomic**; the noble gases, helium, He, neon, Ne, argon, Ar, etc. are **monatomic**.

6.2 Worked Examples

Example 6.1

Ethene reacts with oxygen according to the equation:

$$C_2H_4(g) + 3O_2(g) \rightarrow 2CO_2(g) + 2H_2O(g)$$

15.0 cm^3 of ethene were mixed with 45.0 cm^3 of oxygen and the mixture was sparked to complete the reaction. If all volumes were measured at a pressure of one atmosphere and a temperature of 120°C, the volume of the products would be

A 30 cm^3
B 45 cm^3
C 60 cm^3
D 75 cm^3
E 90 cm^3

(NISEC)

The answer is **C**.

$$C_2H_4(g) + 3O_2(g) = 2CO_2(g) + 2H_2O(g)$$

	1 molecule	3 molecules	2 molecules	2 molecules
By Avogadro's principle	1 volume	3 volumes	2 volumes	2 volumes
	15 cm^3	45 cm^3	30 cm^3	30 cm^3

$$\underbrace{\qquad\qquad}_{60 \text{ cm}^3}$$

Note Usually volumes are measured at *room* temperature and pressure, in which case the water would condense and its volume (though not the number of molecules) would be *negligible*. Remember, Avogadro's principle applies *only* to gases, not to liquids or solids.

Example 6.2

What is the minimum mass of sodium hydrogencarbonate which on heating will produce 3 dm^3 of carbon dioxide measured at room temperature and pressure? (H = 1, C = 12, O = 16, Na = 23 and one mole of carbon dioxide at room temperature and pressure occupies 24.0 dm^3.) The equation for the reaction is

$$2NaHCO_3 = Na_2CO_3 + H_2O + CO_2$$

A 7 g
B 14 g
C 21 g
D 84 g
E 168 g

The answer is C.

(NISEC)

$$2NaHCO_3 = Na_2CO_3 + H_2O + CO_2$$

Reacting masses/volumes   2 × 84 g   24 dm^3

∴ 24 dm^3 CO$_2$ produced from 168 g NaHCO$_3$

∴ 3 dm^3 CO$_2$ produced from $\dfrac{168}{24} \times 3$ g NaHCO$_3$

= 21 g NaHCO$_3$

Example 6.3

Which of the gaseous hydrocarbons listed below would burn in oxygen to produce a mixture of carbon dioxide and steam having a volume equal to three times the volume of the original hydrocarbon? (All volumes measured at the same temperature and pressure.)

A C_2H_4
B CH_4
C C_2H_6
D C_3H_6

The answer is B.

Since the relative numbers of molecules in the reactions are the same as the relative volumes of the gases, by Avogadro's principle:

1 volume of C_2H_4 produces 2 volumes of CO_2 + 2 volumes of steam
1 volume of CH_4 produces 1 volume of CO_2 + 2 volumes of steam
1 volume of C_2H_6 produces 2 volumes of CO_2 + 3 volumes of steam
1 volume of C_3H_6 produces 3 volumes of CO_2 and 3 volumes of steam

6.3 Self-test Questions

Question 6.1

100 cm³ of nitrogen oxide gas (NO) combine with 50 cm³ of oxygen to form 100 cm³ of a single gaseous compound, all volumes being measured at the same temperature and pressure. Which of the following equations fits these facts?

A $NO(g) + O_2(g) \rightarrow NO_3(g)$
B $NO(g) + 2O_2(g) \rightarrow NO_5(g)$
C $2NO(g) + O_2(g) \rightarrow 2NO_2(g)$
D $2NO(g) + O_2(g) \rightarrow N_2O_4(g)$
E $2NO(g) + 2O_2(g) \rightarrow N_2O_6(g)$

(L)

Question 6.2

A sample of gas, mass 0.39 g, was enclosed in a gas syringe and the volume of the gas was recorded at various temperatures but at a constant pressure of 1 atm; the results are shown in the table below.

Temperature (°C)	0	40	60	80	100
Volume (cm³)	224	258	274	291	308

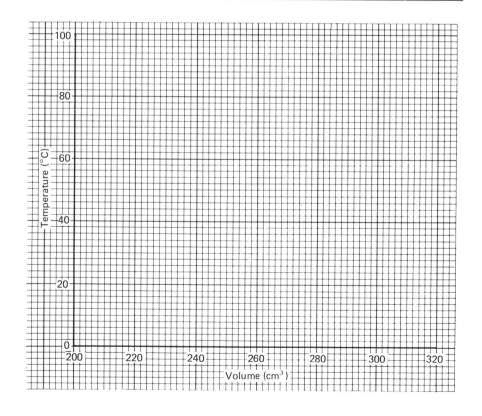

(a) Plot the data on the grid provided and draw the best straight line through the points.
(3 marks)

(b) Read from the graph the volume of the gas at 20°C. **(2 marks)**

(c) Calculate the mass of gas which would have a volume of 24 100 cm^3 at 20°C. **(2 marks)**

(d) What is the relative molecular mass of the gas to the nearest whole number? **(1 mark)**

Question 6.3

What volume of hydrogen is produced at room temperature and pressure, when 0.6 g of magnesium reacts with excess 1 M sulphuric acid?

$$Mg(s) + 2H^+(aq) \rightarrow Mg^{2+}(aq) + H_2(g)$$

(Relative atomic mass: Mg = 24; 1 mole of any gas occupies 24 000 cm^3 at room temperature and pressure.)

A 240 cm^3

B 600 cm^3

C 1000 cm^3

D 14 400 cm^3

E 24 000 cm^3 (L)

Question 6.4

Are the following statements true or false?

(i) The volume of a fixed mass of gas increases when the temperature is increased and decreases when the pressure is increased.

(ii) The volume of 1 mol of hydrogen molecules is twice the volume of 1 mol of helium atoms at the same temperature and pressure.

6.4 Answers to Self-test Questions

6.1 **C.**

From Avogadro's principle, the ratio of reacting molecules is the same as that of reacting volumes, i.e. 2 : 1 : 2.

6.2 (a)

The graph should be a single, neat straight line, cutting the horizontal axis at 224 cm^3 and with the points clearly and accurately marked.

(b) 241 cm^3.

(c) 241 cm^3 of gas at 20°C and 1 atm has a mass of 0.39 g.

∴ 24 100 cm^3 of gas at 20°C and 1 atm has a mass of 39 g.

(d) 39

---

24 000 cm^3 of gas at r.t.p. is 1 mol of the gas.

6.3 B

Reacting masses/volumes

$$Mg(s) + 2H^+(aq) \rightarrow Mg^{2+}(aq) + H_2(g)$$

24 g   24 000 cm^3

24 g Mg produces 24 000 cm^3 H$_2$

∴ 0.6 g Mg produces $\dfrac{24\,000}{24} \times 0.6$ cm^3 H$_2$

= 600 cm^3 H$_2$

6.4 (i) True   (ii) False

The molar volume is the *same* for all gases at the same temperature and pressure.

---

7 Electrochemistry

7.1 Introduction

Substances that allow electricity to pass through them are known as **conductors**: those that do not are called **insulators**.

Metallic elements conduct electricity when solid or molten but are not decomposed by it (compare with electrolytes below). The current is carried by loosely held electrons, which move from one atom to the next.

Non-metallic elements do not conduct electricity because their electrons cannot move from one atom to the next (graphite is an exception).

An **electrolyte** is a compound which, when molten and/or in solution, conducts an electric current and is decomposed by it (compare with metals above). The current is carried by ions, which are free to move through the liquid to the oppositely charged electrodes where they are discharged. Acids, alkalis and salts are examples of electrolytes.

A **non-electrolyte** is a compound which does not conduct an electric current, either when molten or in solution, because no ions are present. Sugar and ethanol are non-electrolytes. Most metal compounds, particularly salts, have giant ionic lattices (see Section 4.2). When they are dissolved in water the ions separate and diffuse throughout the liquid, thus providing a solution which is a good conductor of electricity. Such substances are examples of *strong electrolytes*.

Covalent substances are generally non-electrolytes. However, when the covalent gas, hydrogen chloride, dissolves in water, it forms hydrochloric acid, which is a strong electrolyte. The water reacts with the gas molecules, changing them into ions.

$$HCl(g) + H_2O(l) \rightarrow H_3O^+(aq) + Cl^-(aq)$$

Sulphuric acid and nitric acid molecules also react with water in this way.

Some other covalent substances, such as ammonia and ethanoic acid, have similar reactions with water but form far fewer ions. They are examples of *weak electrolytes*, their solutions being poor conductors of electricity, e.g.

$$NH_3(aq) + H_2O(l) \rightleftharpoons NH_4^+(aq) + OH^-(aq)$$

7.2 Electrolysis

Electrolysis is the decomposition of an electrolyte by the passage of an electric current through it.

During electrolysis negatively charged ions are attracted to the **anode** (positive electrode) and so are called **anions**; positively charged ions are attracted to the **cathode** (negative electrode) and are called **cations**. A useful mnemonic is:

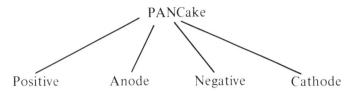

Since electrons are removed at the anode, the process taking place is oxidation; at the cathode, electrons are added and so reduction is occurring (see Section 4.3).

(a) The Electrolysis of Lead(II) Bromide (Fig. 7.1)

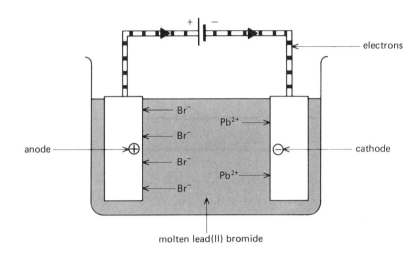

Fig. 7.1

Solid lead(II) bromide does not conduct electricity because its ions are fixed in the crystal lattice. When the solid is melted, the ions are free to move and travel to the oppositely charged electrodes. At the anode, electrons are removed from the bromide ions giving bromine atoms, which combine in pairs to form molecules; reddish brown bromine gas is given off. At the cathode, electrons are added to the lead ions, forming lead atoms, and a bead of molten lead collects at the bottom of the cell.

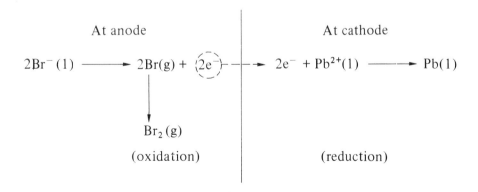

(b) The Electrolysis of Aqueous Solutions

Water is a very weak electrolyte; one molecule in about 550 million ionises as shown:

$$H_2O(l) \rightleftharpoons H^+(aq) + OH^-(aq)$$

61

Consider the electrolysis of dilute sulphuric acid in the apparatus shown in Fig. 7.2. The following ions are present.

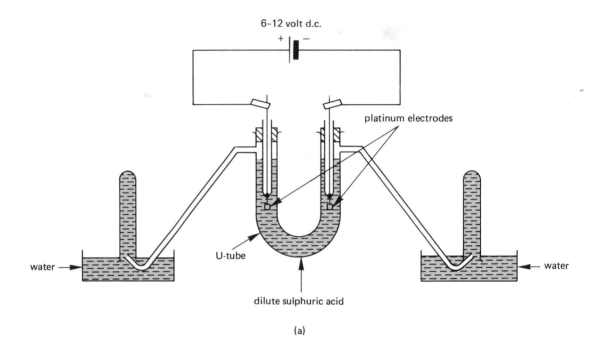

platinum electrodes

U-tube

water →

← water

dilute sulphuric acid

(a)

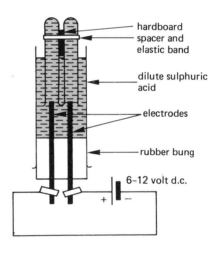

hardboard
spacer and
elastic band

dilute sulphuric
acid

electrodes

rubber bung

6–12 volt d.c.

(b)

Fig. 7.2

	Ions attracted to anode	Ions attracted to cathode
From sulphuric acid	SO_4^{2-} (aq)	$2H^+$ (aq)
From water (very few ions)	OH^- (aq)	H^+ (aq)

At the anode, hydroxide ions are discharged in preference to sulphate ions, forming water and oxygen: hydrogen ions are discharged at the cathode, giving hydrogen gas.

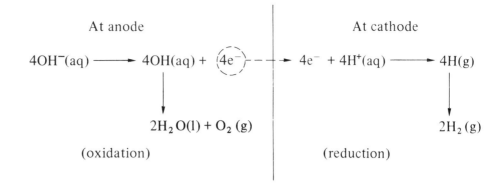

At anode   At cathode

$4OH^-(aq) \longrightarrow 4OH(aq) + (4e^-)---|-\rightarrow 4e^- + 4H^+(aq) \longrightarrow 4H(g)$

$2H_2O(l) + O_2(g)$ $2H_2(g)$

(oxidation)   (reduction)

The equations show why the volume of hydrogen collected is found to be twice that of the oxygen (Avogadro's principle – see Section 6.1).

(c) **Predicting the Products of the Electrolysis of Aqueous Solutions**

First, all of the ions in the solution should be listed under the electrodes to which they are attracted, as in the above example. The ions that will be discharged may, in most cases, be predicted from two simple rules:

1. At the anode, hydroxide ions will be discharged and the products will be oxygen and water, unless:
   (a) chloride, bromide or iodide ions are present (see Example 7.4); or
   (b) the anode dissolves or reacts with the oxygen formed from the hydroxide ions (see Examples 7.1 and 7.3).
2. At the cathode, the ions of the element which is lowest in the reactivity series (see Section 13.1) will be discharged unless the cathode is made of mercury (see Example 18.4 and Question 7.1).

(d) **Applications of Electrolysis**

1. The manufacture of elements. Examples include sodium, aluminium and chlorine (see Section 18.2).
2. The manufacture of compounds, e.g. sodium hydroxide (see Section 18.3).
3. Purification of metals. Copper is purified by electrolysing aqueous copper(II) sulphate solution with blocks of impure copper as anodes and thin sheets of the pure metal as cathodes. Pure copper is transferred from the anode to the cathode (see Example 7.3).
4. Electroplating. The article to be plated is made the cathode in the electrolysis cell, the anode is made of the plating metal and the electrolyte is a solution containing its ions. When a current passes, the plating metal is transferred from the anode to the cathode, as in the purification of copper.
5. Anodising. Aluminium is coated with a layer of aluminium oxide which protects the metal from corrosion. This oxide coating can be thickened by making the aluminium the anode during the electrolysis of sulphuric acid. The oxide layer will absorb dyes and so is often coloured.

7.3 Worked Examples

Example 7.1

(a) From the following list of substances: zinc bromide, paraffin wax, hydrogen chloride, copper, potassium chloride and iodine, name those that:

(i) conduct electricity in the solid state,

Copper.

(ii) conduct electricity when molten,

Zinc bromide, copper, potassium chloride.

(iii) conduct electricity in aqueous solution,

Zinc bromide, hydrogen chloride, potassium chloride.

(iv) do not readily conduct under any conditions.   **(4 marks)**

Paraffin wax, iodine.

Check with Section 7.1 if you are uncertain about these answers.

(b) Draw a simple circuit (Fig. 7.3) to show that the substance you chose in (a) (i) will conduct an electric current.   **(2 marks)**

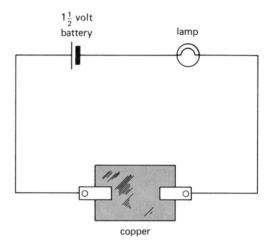

Fig. 7.3

(c) The electrolysis of an aqueous solution of calcium hydroxide was investigated using carbon electrodes.

(i) Which gas would you expect to be produced at the anode?

Oxygen.

Only OH⁻ ions (from both the calcium hydroxide and the water) are present at the anode. Therefore the electrode equation is

$$4OH^-(aq) \longrightarrow 4OH(aq) + 4e^-$$
$$\downarrow$$
$$2H_2O(l) + O_2(g)$$

(ii) During the electrolysis the anode decreases in mass and the solution around it becomes milky. Explain these observations.   **(3 marks)**

The oxygen produced reacts with the carbon electrode to give carbon dioxide which will turn calcium hydroxide solution (limewater) milky, owing to the formation of insoluble calcium carbonate.

(d) The apparatus was set up as shown in Fig. 7.4 and after some time a yellow substance was produced in the position shown.

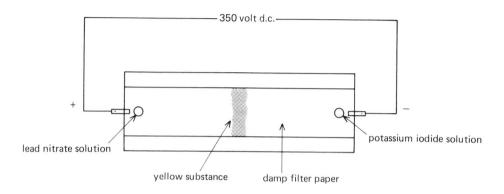

Fig. 7.4

(i) Give the formulae of the ions present in the lead nitrate and potassium iodide solutions.

From $Pb(NO_3)_2$ $Pb^{2+}(aq)$ *and* $NO_3^-(aq)$
From KI $K^+(aq)$ *and* $I^-(aq)$
From water $H^+(aq)$ *and* $OH^-(aq)$

(ii) What is the yellow substance? Write an ionic equation to show its formation.
Lead iodide
$$Pb^{2+}(aq) + 2I^-(aq) \rightarrow PbI_2(s)$$

(iii) Explain how the yellow compound is formed.   **(6 marks)**

The lead ions are attracted to the cathode, and the iodide ions are attracted to the anode. Where they meet, insoluble lead(II) iodide is formed.

Example 7.2

(a) Potassium iodide will conduct an electric current when molten. Name the product formed at
    (i) the negative electrode,   **(1 mark)**
Potassium.

    (ii) the positive electrode.   **(1 mark)**
Iodine.

(b) Why does potassium iodide not conduct electricity in the solid state?   **(1 mark)**
In the solid state, the ions are not free to move.

Example 7.3

(a) Copper may be purified by electrolysis of aqueous copper(II) sulphate solution.
    (i) Give the name, charge and material of each electrode.   **(3 marks)**
         Name: *Anode*   Name: *Cathode*
         Charge: *Positive*   Charge: *Negative*
         Material: *Impure copper*   Material: *Thin sheet of pure copper*

65

(ii) Describe briefly how the copper is purified. **(2 marks)**

At the anode, electrons are removed from copper atoms, forming Cu^{2+} ions which go into solution. Impurities dissolve or sink to the bottom of the container. At the cathode, Cu^{2+} ions gain electrons, forming copper atoms, and so the cathode becomes coated with pure copper.

(iii) Write ion-electron half equations for the reactions that take place at each electrode.

(2 marks)

At anode   *At cathode*

$$Cu(s) \longrightarrow Cu^{2+}(aq) + 2e^- \quad | \quad \longrightarrow 2e^- + Cu^{2+}(aq) \longrightarrow Cu(s)$$

(iv) Explain which reaction is an oxidation. **(1 mark)**

The reaction at the anode is oxidation because each copper atom loses two electrons here.

	Ions attracted to anode	*Ions attracted to cathode*
From $CuSO_4$	SO_4^{2-} (aq)	Cu^{2+} (aq)
From H_2O (very few ions)	OH^- (aq)	H^+ (aq)

No gases are given off during this electrolysis. At the anode it requires less energy to remove electrons from the copper atoms of the anode itself than to remove them from the OH^- or SO_4^{2-} ions. At the cathode Cu^{2+} ions are discharged in preference to H^+ ions because copper is below hydrogen in the reactivity series.

(b) Aqueous copper(II) sulphate may be regarded as an *electrolyte*. Explain why solid copper is not regarded as an electrolyte although it conducts electricity. **(4 marks)**

Electrolytes conduct electricity only when molten or in solution, and are decomposed by it. The current is carried by ions. In solid copper the current is carried by loosely held electrons and the metal remains chemically unchanged. Therefore it is not an electrolyte.

(c) Explain why it is not possible to purify aluminium in a similar way. **(2 marks)**
(OLE)

Hydrogen ions and aluminium ions would both be present at the cathode but the hydrogen ions would be discharged, since hydrogen is below aluminium in the reactivity series (see Section 7.2). The aluminium ions would remain in solution.

Example 7.4

(a) Name the particles responsible for carrying the current in (i) a wire, and in (ii) an electrolyte. **(2 marks)**

(i) Electrons.

(ii) Ions.

The apparatus shown in Fig. 7.5 was assembled and the circuit completed. Immediately bubbles of gas were seen at the cathode (negative electrode), but there was little apparent reaction at the anode. After some time it was noticed that the solution around the cathode had turned blue while the area near the anode was colourless: there were now some bubbles at the anode.

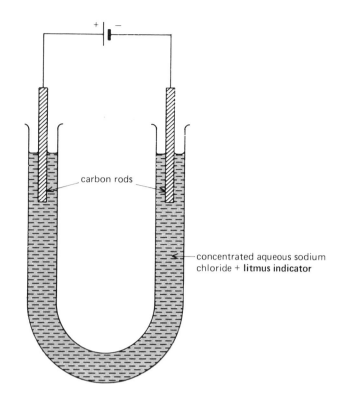

Fig. 7.5

(b) (i) Name the gas evolved at the **anode** and write an ionic equation to show how it was formed.

Chlorine is evolved at the anode:

$2Cl^-(aq) \rightarrow 2Cl(g) + 2e^-$

$\downarrow$

$Cl_2(g)$

	Ions attracted to anode	Ions attracted to cathode
From NaCl	Cl^- (aq)	Na^+ (aq)
From H_2O (very few ions)	OH^- (aq)	H^+ (aq)

Chloride ions are always discharged in preference to hydroxide ions from concentrated solutions (see Section 7.2).

(ii) Why were bubbles of this gas not observed immediately when the circuit was completed?

Chlorine is moderately soluble in aqueous sodium chloride solution and so at first the gas dissolves.

(iii) Explain briefly the reason why the litmus indicator around the anode became colourless.   **(8 marks)**

Chlorine will bleach coloured compounds by oxidising them to colourless products (see Example 17.6).

(c) (i) Name the gas evolved at the **cathode** and write an ionic equation to show how it was formed. Give **one** test to identify this gas.

The gas is hydrogen.

$$2e^- + 2H^+(aq) \rightarrow 2H(g)$$
$$\downarrow$$
$$H_2(g)$$

At the cathode, hydrogen ions are discharged in preference to sodium ions since hydrogen is much lower in the reactivity series than sodium.

Hydrogen could be identified by collecting a tubeful of gas and applying a flame: a 'pop' could be clearly heard.

(ii) Explain why the litmus solution around the cathode turned blue. **(8 marks)**

The removal of hydrogen ions from the water molecules around the cathode leaves an excess of hydroxide ions and these turn the litmus blue.

(d) Name the sodium compound which is manufactured by the electrolysis of brine, and describe briefly its importance in the chemistry of soap production. **(6 marks)**

(AEB, 1981)

Sodium hydroxide is manufactured by the electrolysis of brine (see Example 18.4). Animal fats and vegetable oils are esters of long-chained carboxylic acids and propane-1,2,3-triol. When boiled with sodium hydroxide solution, they are hydrolysed to propane-1,2,3-triol and the sodium salt of the acid, which is soap.

Example 7.5

The table gives information about 5 substances.

Substance	Melting point (°C)	Does it conduct electricity when solid?	Does it conduct electricity when molten?
A	800	No	Yes
B	3700	Yes	Yes
C	−39	Yes	Yes
D	0	No	No
E	−117	No	No

Which letter in the table corresponds to each substance given below?
(a) ethanol   *E*
(b) mercury   *C*
(c) graphite   *B*
(d) sodium chloride   *A*
(e) water   *D*   **(5 marks)**

Only mercury and graphite will conduct electricity in the solid state, and the melting point suggests that **C** is mercury. Sodium chloride, being the only ionic solid, will conduct electricity when molten but not when solid, and so **A** must be sodium chloride. Water and ethanol are covalent and non-conductors of electricity, and the melting point suggests that **D** is water rather than ethanol.

7.4 Self-test Questions

Question 7.1

A concentrated solution of copper(II) chloride was electrolysed using inert electrodes.

After 10 min, a pink substance was deposited on one electrode and a gas which bleached moist universal indicator paper was evolved at the other electrode.

After 1 hour, the same products were being formed but the solution was paler in colour.

After 2 hours, the pink substance was no longer being deposited: instead, a colourless gas was evolved. At the other electrode, the bubbles of gas no longer bleached moist universal indicator paper.

After 3 hours, the liquid could no longer conduct electricity.

(a) Give the formulae of all ions present in the copper(II) chloride solution.   (2 marks)
(b) (i) At which electrode was the pink substance deposited?
    (ii) Which gas is first discharged at the other electrode?   (2 marks)
(c) Why did the colour of the copper(II) chloride solution fade?   (2 marks)
(d) Name the two gases being evolved after two hours. Write ion–electron equations for each reaction and describe the tests you would carry out to confirm the identity of each gas.   (6 marks)
(e) Why did the electrolysis stop after 3 hours?   (1 mark)

Questions 7.2 and 7.3

Directions. These two questions deal with laboratory situations. Each situation is followed by a set of questions. Select the best answer for each question.

The apparatus below was used to investigate the electrolysis of sodium chloride, NaCl. The crucible contained molten sodium chloride. The U-tube contained an aqueous solution of sodium chloride. The bulb lit when the switch was closed.

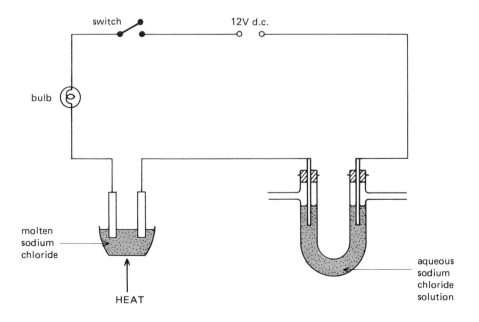

Question 7.2

What were the products at the two cathodes in this experiment?

	At the cathode in molten sodium chloride	*At the cathode in aqueous sodium chloride solution*
A	Sodium	Hydrogen
B	Chlorine	Oxygen
C	Hydrogen	Hydrogen
D	Sodium	Sodium
E	Chlorine	Chlorine

(L)

Question 7.3

What would happen in the experiment described on the previous page if the heat supply under the crucible were to be removed and the contents allowed to cool to room temperature?

	Bulb	*Electrolysis in the crucible*	*Electrolysis in the U-tube*
A	Stays alight	Continues	Continues
B	Stays alight	Stops	Continues
C	Stays alight	Continues	Stops
D	Goes out	Stops	Continues
E	Goes out	Stops	Stops

(L)

Question 7.4

In the refining of copper, the reaction occurring at the negative electrode is given by:

A $Cu^{2+} + 2e \rightarrow Cu$
B $Cu^{2+} + 2e \rightarrow 2Cu$
C $Cu - 2e \rightarrow Cu^{2+}$
D $4OH^- - 4e \rightarrow 2H_2O + O_2$
E $Cu^+ + e \rightarrow Cu$

7.5 Answers to Self-test Questions

7.1 (a)

	Ions attracted to anode	*Ions attracted to cathode*
From $CuCl_2$	Cl^- (aq)	Cu^{2+} (aq)
From H_2O (very few ions)	OH^- (aq)	H^+ (aq)

(b) (i) The pink deposit is copper formed at the cathode.
(ii) Chlorine.
(c) The copper ions are responsible for the colour of the solution. These ions are steadily removed as copper atoms and so the colour becomes paler.
(d) The products are oxygen (relights a glowing splint) and hydrogen (pops in a flame).

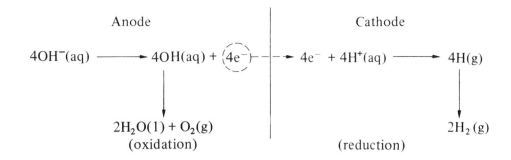

Anode

$$4OH^-(aq) \longrightarrow 4OH(aq) + 4e^-$$

$$2H_2O(l) + O_2(g)$$
(oxidation)

Cathode

$$4e^- + 4H^+(aq) \longrightarrow 4H(g)$$

$$2H_2(g)$$
(reduction)

After 2 hours, there are very few copper or chloride ions left in solution, so hydrogen ions are discharged at the cathode and hydroxide ions are discharged at the anode.

(e) After 3 hours, there are no copper or chloride ions left to carry the current. Pure water is a very poor conductor of electricity because hardly any ions are present.

7.2 **A**.

Since hydrogen is below sodium in the reactivity series.

7.3 **E**.

Sodium chloride solidifies; ions can no longer travel to the electrodes.

7.4 **A**.

8 Energy Changes in Chemistry

8.1 Introduction

Energy is usually defined as the capacity for doing work. It can be neither created nor destroyed (Law of conservation of energy) but it can be converted from one form to another. For example, the chemical energy stored up in petrol can be converted to heat energy and kinetic energy (energy associated with movement) in a car engine. Another important energy conversion is the changing of solar energy (energy from the sun) into chemical energy in photosynthesis (see Section 10.4).

Most of the energy which we use today is obtained from 'fossil fuels' — gas, oil and coal. Supplies of these fuels will not last forever, and they must not be wasted. They can be supplemented by nuclear energy (see Section 2.5), and energy from wind, flowing water and from the sun, but these last three energy sources do not contribute very greatly to world supplies at the moment. They are, however, renewable: they can be used again and again, unlike fossil and nuclear fuels. Also, they do not cause pollution. Combustion of fossil fuels causes air pollution (see Section 10.1), and nuclear reactors give rise to radioactive waste. Choice of fuel depends largely upon availability, convenience and cost. Only fairly recently has large-scale concern about pollution had any bearing on the matter.

During almost all chemical changes energy is exchanged with the surroundings: if it is given out by the reaction, the change is said to be *exothermic*; if it is absorbed, the reaction is *endothermic*. The reason that the energy changes occur is that bonds between atoms or ions in the reactants have to be broken and new ones formed in the products. Bond breaking absorbs energy but bond making releases it. The overall energy change that occurs results from the difference between the energy supplied for the breaking of the reactant bonds and that evolved in the making of the product bonds.

The symbol for heat change occurring at constant pressure is ΔH, where Δ (delta) means 'change of' and H is the 'heat content' or *enthalpy* of the system. ΔH is *negative* for an exothermic change because the system *loses* energy to the surroundings: in an endothermic reaction ΔH is *positive* because the system *gains* energy from the surroundings. These enthalpy changes are often simply called 'heat changes'.

The unit of energy is the joule (J) or, more conveniently, the kilojoule (kJ), which is 1000 joules. One **joule** is the energy required to raise the temperature of 1 g of water by

$$\frac{1}{4.18} \text{ K } (\frac{1}{4.18}°C);$$

thus 4.18 joules will raise the temperature of 1 g of water by 1 K (1°C).

The **enthalpy change of reaction (heat of reaction)**, ΔH, is the heat change that occurs when the numbers of moles of reactants indicated by the equation react together. This information is often included in the equation. For instance

$$C(s) + O_2(g) \rightarrow CO_2(g) \qquad \Delta H = -394 \text{ kJ mol}^{-1}$$

means that when 1 mol of carbon burns in excess oxygen, the heat *evolved* is 394 kJ.

8.2 Measurement of Enthalpy Changes (Heat Changes)

The **enthalpy change (heat) of combustion** of a substance is the heat change that occurs when one mole of it is completely burned in excess oxygen.

The enthalpy change of combustion of a liquid such as ethanol can be measured by burning a *known* mass of it and using the heat produced to raise the temperature of a *known* mass of water in the apparatus shown in Fig. 8.1.

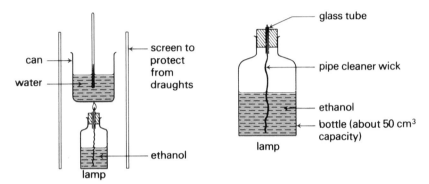

Fig. 8.1

The **enthalpy change (heat) of neutralisation** is the heat change that occurs when one mole of hydrated hydrogen ions from an acid reacts with one mole of hydrated hydroxide ions from an alkali.

The apparatus of Fig. 8.2 may be used to measure the temperature rise when a known volume of acid, concentration also known, is neutralised by a known volume of alkali. The enthalpy change of neutralisation is calculated from the result. Heat losses in this case are negligible (see Example 8.4).

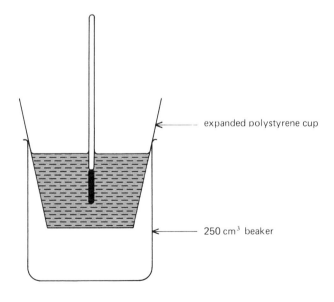

Fig. 8.2

8.3 Electrical Cells

Electrical cells enable electrical energy to be obtained from chemical reactions. This is the reverse of what happens during electrolysis, where chemical reactions are brought about by the use of electrical energy. The simplest type of cell consists of two metals dipping into an electrolyte (see Section 7.1) and connected to one another by a wire.

Suppose that the metals are zinc and copper and the electrolyte is dilute sulphuric acid (Fig. 8.3). Zinc, being more reactive than copper, forms ions more readily and so its atoms each give up two electrons and go into solution as Zn^{2+} ions. The electrons flow through the wire to the copper rod where H^+ ions from the acid take them to become H atoms, which join in pairs to form H_2 molecules.

At zinc $\qquad\qquad Zn(s) \rightarrow Zn^{2+}(aq) + 2e^-$

At copper $\quad 2H^+(aq) + 2e^- \rightarrow H_2(g)$

Adding: $\quad 2H^+(aq) + Zn(s) \rightarrow Zn^{2+}(aq) + H_2(g) \qquad$ (overall change)

Clearly, the overall chemical change taking place is the same as that which occurs when zinc is dipped into dilute sulphuric acid, i.e. displacement, but the cell enables us to obtain electrical energy from the reaction.

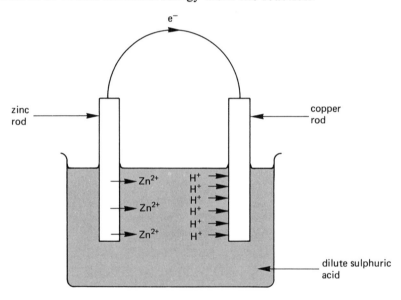

Fig. 8.3

8.4 Light Energy in Chemistry

Light energy can be produced by a chemical reaction (e.g. in a flame) or can be absorbed (e.g. in the reaction between hydrogen and chlorine (see Example 17.3), in photography and, very importantly, in photosynthesis (see Section 10.4)).

8.5 Worked Examples

Examples 8.1 and 8.2 concern the following graphs of a quantity Y plotted against a quantity X.

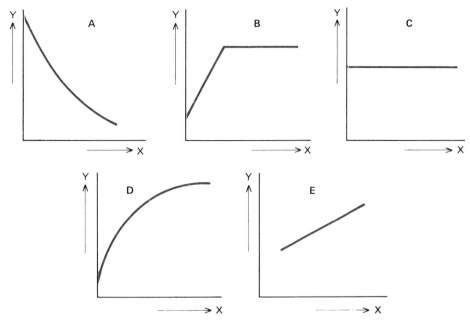

Fig. 8.4

Select, from A to E in Fig. 8.4, the graph that best represents the way Y changes with X, when Y and X are the quantities shown in each question below.

Example 8.1

100 cm^3 of water, at room temperature, was placed in a beaker and then heated with a burner supplying heat at a constant rate. The temperature of the water was recorded at regular intervals until 10 cm^3 of it had boiled away.

Y = Temperature of water   X = Time.   (L)

The answer is **B.**

The temperature rises steadily until the liquid boils and then remains constant.

Example 8.2

A series of experiments was carried out to measure the heat given out when each of the following hydrocarbons was burned.
CH_4, C_2H_6, C_3H_8, C_4H_{10}, C_5H_{12}
Y = Heat given out on the   X = Relative molecular mass
    combustion of 1 mole   of hydrocarbon
    of hydrocarbon   (L)

The answer is **E.**

Each hydrocarbon differs from the next by the group $-CH_2-$, so the relative molecular mass increases steadily. As 1 mole of each successive member burns, more energy must be *supplied* to break an extra 1 mole of C−C bonds and 2 moles of C−H bonds. However, energy will be *evolved* on making the bonds in an extra 1 mole of carbon dioxide and 1 mole of water. The difference between the total energy absorbed and the total energy evolved will thus change by the same amount on going from one hydrocarbon to the next, giving graph E.

Example 8.3

The enthalpy changes of combustion of a number of cycloalkanes are as follows:
cyclopentane C_5H_{10} $\Delta H = -3317$ kJ mol^{-1}
cyclohexane C_6H_{12} $\Delta H = -3948$ kJ mol^{-1}
cycloheptane C_7H_{14} $\Delta H = -4632$ kJ mol^{-1}
When the enthalpy change of combustion of cyclononane, C_9H_{18}, is measured, the following data are obtained: the combustion of 0.21 g of cyclononane produces 9.96 kJ of heat.

(a) Calculate the enthalpy change of combustion of cyclononane. **(4 marks)**

Mol of C_9H_{18} $= mass/molar\ mass\ (see\ Section\ 5.1)$
Molar mass of C_9H_{18} $= (9 \times 12) + (18 \times 1) = 126\ g\ mol^{-1}$
$\therefore$ Mol of C_9H_{18} $= 0.21/126 = 1/600$
If 1/600 mol of cyclononane produces 9.96 kJ
1 mol of cyclononane produces $9.96 \times 600 = 5976$ kJ
$\therefore$ Enthalpy change of combustion of cyclononane is -5976 kJ mol^{-1}

Note the minus sign: heat is given out on burning.

(b) Plot a graph of enthalpy change of combustion of the cycloalkanes (on the y axis) (see Fig. 8.5) against number of carbon atoms in their molecules (on the x axis). **(3 marks)**

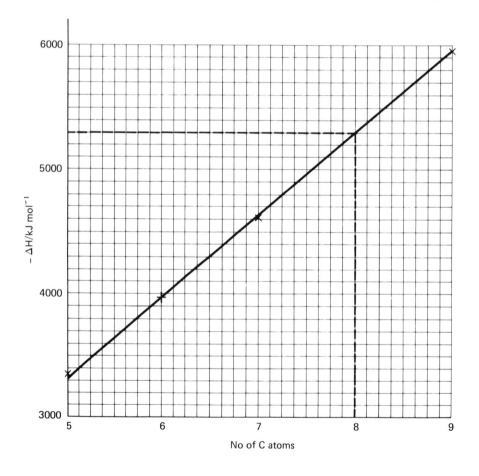

Fig. 8.5

(c) An unknown cycloalkane has an enthalpy change of combustion of -5305 kJ mol^{-1}. Use the graph to find out how many carbon atoms this cycloalkane contains. **(2 marks)**

The cycloalkane contains 8 carbon atoms (Fig. 8.5).

(d) What is the formula of the unknown cycloalkane? **(1 mark)**

The formula is C_8H_{16}.

The general formula of this series is C_nH_{2n}. In this case $n = 8$, so $2n = 16$.

(e) Write an equation to represent its combustion to carbon dioxide and water. **(2 marks)**

$C_8H_{16} + 12O_2 \rightarrow 8CO_2 + 8H_2O$

Example 8.4

40 g of sodium hydroxide was dissolved in water so as to make 500 cm^3 of solution. 50 cm^3 portions of this solution were placed in insulated plastic cups and the temperature of each was noted. Various quantities of aqueous hydrochloric acid (at the same temperature) were added to each cup, and the temperature rise was observed; hence the heat change in kJ was calculated. The results are shown in Table 1.

Table 1

Volume of aqueous sodium hydroxide (cm^3)	Volume of aqueous hydrochloric acid (cm^3)	Heat energy evolved (kJ)
50	10	1.1
50	20	2.2
50	30	3.4
50	40	4.5
50	50	5.6
50	60	5.6
50	70	5.6

(a) Explain briefly why the heat change becomes constant after 50 cm^3 has been added. **(2 marks)**

All of the sodium hydroxide solution must be neutralised when 50 cm^3 of aqueous hydrochloric acid is added. The extra acid added in the last two experiments has no alkali to react with and therefore produces no extra heat.

(b) How many moles of sodium hydroxide are contained in 50 cm^3 of solution? **(2 marks)**

Molar mass of NaOH = (23 + 16 + 1) = 40 g mol^{-1}
∴ 500 cm^3 of solution contain 1 mol of NaOH
∴ 50 cm^3 of solution contain 0.1 mol of NaOH

(c) Calculate the concentration of the aqueous hydrochloric acid in mol dm^{-3}. **(2 marks)**

The equation for the reaction is
NaOH(aq) + HCl(aq) → NaCl(aq) + H$_2$O(l)
∴ 1 mol of NaOH reacts with 1 mol of HCl
50 cm^3 of NaOH(aq) containing 0.1 mol of NaOH are neutralised by 50 cm^3 of HCl
∴ There must be 0.1 mol of HCl in the 50 cm^3 of HCl(aq)
∴ Concentration of HCl(aq) is $0.1 \times \dfrac{1000}{50} = 2$ mol dm^{-3} (or 2 mol/l)

(d) What quantity of heat energy would be evolved if a solution containing 1 mol of sodium hydroxide were neutralised by aqueous hydrochloric acid? **(2 marks)**

0.1 mol of NaOH evolves 5.6 kJ of heat on neutralisation

∴ 1 mol of NaOH evolves 56 kJ of heat on neutralisation

(e) Estimate the heat energy which would be evolved if a solution containing 1 mol of a metal hydroxide, $M(OH)_2$, was neutralised by aqueous hydrochloric acid. Explain briefly the reasoning behind your answer, and write both a molecular equation and an ionic equation for the reaction. **(4 marks)**

Since there are 2 mol of OH^- ions to be neutralised, twice as much heat energy will be evolved, i.e. 112 kJ. The M^{2+} ions and Cl^- ions do not take part in the reaction.

Equation $M(OH)_2(aq) + 2HCl(aq) \rightarrow MCl_2(aq) + 2H_2O(l)$

Ionic equation $(OH)^-(aq) + H^+(aq) \rightarrow H_2O(l)$

(f) Describe briefly how a reasonably pure crystalline sample of the hydrated salt formed in (e) could be isolated from its solution. **(4 marks)**

(AEB, 1982)

See Section 1.3(a) (ii).

Example 8.5

One mole of magnesium atoms reacts with exactly one mole of copper(II) ions. In a series of experiments magnesium powder was added to copper(II) sulphate solution. In which of the following experiments would the temperature rise be the greatest? (Relative atomic mass of Mg = 24.)

A 9.6 g of Mg added to 1 dm^3 of 0.1 M aqueous $CuSO_4$

B 4.8 g of Mg added to 1 dm^3 of 0.2 M aqueous $CuSO_4$

C 2.4 g of Mg added to 1 dm^3 of 0.2 M aqueous $CuSO_4$

D 4.8 g of Mg added to 1 dm^3 of 0.1 M aqueous $CuSO_4$

E 2.4 g of Mg added to 1 dm^3 of 0.4 M aqueous $CuSO_4$ (AEB, 1983)

*The answer is **B**.*

The Mg atoms and Cu^{2+} ions react in the ratio 1 : 1 and in B there are 0.2 mol of Mg and 0.2 mol of Cu^{2+}. None of the other mixtures contains as much *reacting* material. For example, A contains 0.4 mol of Mg but only 0.1 mol of Cu^{2+}, so only 0.1 mol of Mg can react. The extra Mg makes no difference to the amount of heat produced.

Example 8.6

Using the apparatus shown in Fig. 8.6, 0.8 g of powdered copper is added to 100 cm^3 of 0.2 M silver nitrate solution and the mixture is stirred gently with the thermometer. The temperature at the beginning of the experiment is 18.2°C and it rises to a maximum of 21.7°C. The equation for the reaction is

$2AgNO_3(aq) + Cu(s) \rightarrow Cu(NO_3)_2(aq) + 2Ag(s)$.

(a) Why is a polystyrene cup used instead of a glass container, and why is it placed in the beaker? **(2 marks)**

Expanded polystyrene is a good heat insulator and absorbs hardly any of the heat produced in the reaction. Placing the cup in the beaker improves the insulation by cutting out draughts and makes it less easy to knock over the cup, which is very light.

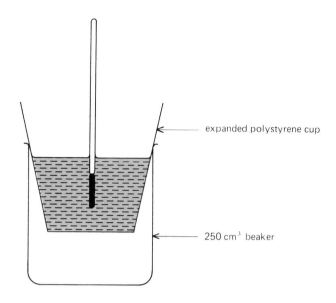

expanded polystyrene cup

250 cm³ beaker

Fig. 8.6

(b) How many moles of silver nitrate are present initially? **(2 marks)**

1 dm³ of 0.2 M silver nitrate solution contains 0.2 mol of AgNO₃
∴ 100 cm³ of the solution contains 0.02 mol of AgNO₃

(c) How many moles of copper atoms are needed to react with the silver nitrate? **(2 marks)**

From the equation, 2 mol AgNO₃ reacts with 1 mol Cu
∴ 0.02 mol AgNO₃ reacts with 0.01 mol Cu

(d) How many moles of copper atoms are present? Is this enough? **(2 marks)**

Moles of Cu = mass/molar mass = 0.8/63.5 = 0.0126
As only 0.01 mol is needed, this is enough.

(e) What is the temperature rise? **(2 marks)**

The temperature rise is (21.7 − 18.2) = 3.5°C

(f) If it takes 4.2 joules to raise the temperature of 1 g of water by 1°C, how much heat is produced in this experiment? **(2 marks)**

Heat produced = mass × 4.2 × temperature rise
*   = 100 × 4.2 × 3.5 J*
*   = 1470 J*

(g) Calculate the enthalpy change of reaction in kJ per mole of copper atoms. **(2 marks)**

0.01 mol of Cu was used, producing 1470 J
∴ For 1 mole of Cu, heat produced would be 147 000 J = 147 kJ
∴ enthalpy of reaction is −147 kJ mol⁻¹.

Note sign.

Example 8.7

In the cell shown in Fig. 8.7 there is a flow of electrons from metal **P** to metal **Q** through the meter.

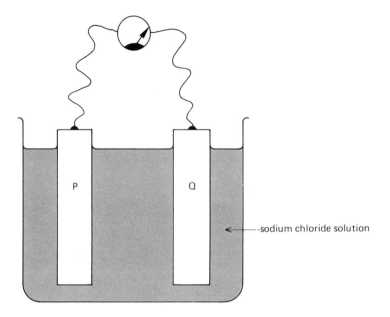

-sodium chloride solution

Fig. 8.7

Which of the following produce this result?

	P	Q
A	silver	copper
B	zinc	copper
C	copper	zinc
D	silver	zinc

The answer is **B.**

(SEB)

Atoms of the more reactive of the two metals give up electrons and go into solution as positive ions. The electrons flow to the less reactive metal (see Section 8.3).

8.6 Self-test Questions

Are the statements in Questions 8.1–8.4 true or false?

Question 8.1

(i) The reaction between methane and chlorine occurs in the presence of ultra-violet light.
(ii) Ultra-violet light is a source of energy which can initiate some chemical reactions.

Question 8.2

(i) The products of an exothermic reaction contain more energy than the reactants at the same temperature and pressure.
(ii) Energy is released during an exothermic reaction.

Question 8.3

    (i) Heat energy is liberated when a fuel is burned in air or oxygen.

    (ii) Energy is liberated when chemical bonds are broken.

Question 8.4

    (i) The melting point of sodium chloride is greater than the melting point of water.

    (ii) The forces of attraction between ions in sodium chloride are greater than the forces between water molecules.

8.7 Answers to Self-test Questions

8.1 (i) True   (ii) True

8.2 (i) False   (ii) True

Since energy is *given out* in an exothermic reaction, the products must contain *less* energy than the reactants.

8.3 (i) True   (ii) False

8.4 (i) True   (ii) True

9 Rate of Reaction and Chemical Equilibrium

9.1   Rate of Reaction

If dilute hydrochloric acid is added to excess marble chips, carbon dioxide is given off and the mass of reactants decreases. A graph of decrease in mass against time is as shown in Fig. 9.1

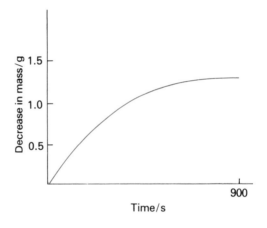

Fig. 9.1

The slope of the graph is steepest at the beginning, showing that the mass is decreasing at the greatest rate (i.e. the rate of evolution of carbon dioxide is greatest) when the acid is at its most concentrated. As the reaction proceeds the slope becomes less steep, indicating that the reaction is slowing down. This is because the acid is being used up and becoming more dilute. Where the graph becomes a horizontal straight line the reaction has ceased: no further change in mass occurs because all of the acid has been used up.

A **catalyst** is a substance which alters the rate of a chemical reaction but may be recovered unchanged in mass at the end. Positive catalysts speed up chemical reactions; negative catalysts (inhibitors) slow them down. Examples of positive catalysts include manganese(IV) oxide in the decomposition of hydrogen peroxide solution into water and oxygen (see Example 9.2), iron in the manufacture of ammonia by the Haber process (see Section 18.5), platinum in the manufacture of nitric acid by the Ostwald process (see Section 18.6) and vanadium(V) oxide in the manufacture of sulphuric acid by the contact process (see Section 18.4). Many biological processes involve natural catalysts, called **enzymes** (see Section 19.5).

The rate of a chemical reaction may be increased by:

1. Making the reactants more finely divided.
2. Increasing the concentration of the reactants, or increasing the pressure if they are gases.
3. Increasing the temperature.
4. Adding a (positive) catalyst.
5. Increasing the light intensity (applies to a few reactions only) (see Example 17.3).

9.2 Reversible Reactions

A **reversible reaction** is one that can proceed in either direction depending on the conditions under which it is carried out.

For example, if blue copper(II) sulphate crystals are heated, water is given off and white anhydrous copper(II) sulphate remains. When water is added to this solid, blue hydrated copper(II) sulphate is formed once more.

$$CuSO_4.5H_2O(s) \rightleftharpoons CuSO_4(s) + 5H_2O(l)$$

(The $\rightleftharpoons$ sign shows that the reaction can proceed in both directions.)

Often a reversible reaction does not go to completion, i.e. the reactants are not all used up. The reason for this is explained in the next section.

9.3 Chemical Equilibrium

If a reversible reaction is carried out in a closed container so that the products cannot escape, the reaction will often appear to stop before all of the reactants are used up. This is because as soon as the products are formed they begin to react together to form the reactants again. At first the forward reaction is much faster than the backward one because the concentrations of the reactants are much greater than the concentrations of the products. As the reaction goes on, the concentrations of the reactants fall, slowing the rate of the forward reaction. At the same time the concentrations of the products are building up, so the backward reaction increases in rate. A point is reached where the two rates are the same and the reaction appears to stop: the system is said to be **in equilibrium**.

9.4 Worked Examples

Example 9.1

This question is about measuring the rate of the reaction between solutions of potassium peroxodisulphate, $K_2S_2O_8$, and potassium iodide, KI. It was studied by the following method: 20 cm³ of potassium peroxodisulphate solution was mixed with 20 cm³ of potassium iodide solution (an excess), 5 cm³ of sodium thiosulphate solution and 1 cm³ of starch solution.

The potassium peroxodisulphate reacts with the potassium iodide at a measurable rate forming iodine. As soon as the iodine is formed it reacts rapidly with the sodium thiosulphate until all the sodium thiosulphate is used up. The iodine formed after the sodium thiosulphate is used up gives a blue colour with the starch. The time taken for the blue colour to appear was recorded.

The experiment was repeated using potassium peroxodisulphate solutions of different concentrations, and the results are shown in the table.

Peroxodisulphate concentration ($mol\ dm^{-3}$)	Time (t) for blue colour to appear (sec)	$\frac{1}{t}$ (sec^{-1})
0.3	20	0.05
0.2	30	0.033
0.15	40	0.025
0.05	120	0.0083

(a) If we wish to investigate the rate of a reaction why do we calculate $1/t$? **(1 mark)**

In each experiment the same mass of iodine is needed to use up the 5 cm³ of sodium thiosulphate solution and then produce the blue colour with the starch. If this constant mass, m, is produced in time t, then the average rate of production from the beginning to time t is m/t. Since m is the same each time we can write:

Average rate $\propto 1/t$

(e.g. if the rate doubles, the time is halved.)

(b) On the axes provided in Fig. 9.2, plot a graph showing how $1/t$ varies with the concentration of the potassium peroxodisulphate solution. **(2 marks)**

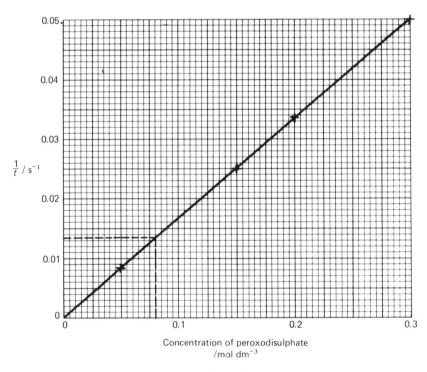

Fig. 9.2

(c) Explain why this graph shows that the rate of the reaction is proportional to the concentration of the potassium peroxodisulphate solution. **(1 mark)**

The straight-line graph indicates direct proportionality between concentration and $1/t$, e.g. when the concentration doubles, trebles, etc., the value of $1/t$ changes by the same factor. Since $1/t$ is proportional to the rate (answer(a)) the rate must change to the same extent as $1/t$ and therefore rate is proportional to concentration.

(d) Use your graph to determine the time that would have been taken for the blue colour to appear if the potassium peroxodisulphate solution had been 0.08 M.   **(1 mark)**

When the concentration is 0.08 M, i.e. 0.08 mol dm^{-3}, $1/t = 0.0135$ s^{-1}
∴ t = 74 s

(e) In all the above experiments the solution of sodium thiosulphate used was 0.01 M. What would have been the time taken for the blue colour to appear if the experiment was repeated using 0.2 M potassium peroxodisulphate solution and 0.02 M sodium thiosulphate solution in the same volume as before?   **(1 mark)**
(L)

The iodine will be produced at the same rate as before but twice as much will be needed to react with the sodium thiosulphate solution before the blue colour appears. Therefore the time taken will be more than 60 seconds.

The reaction slows down as it proceeds, so the time taken to produce twice as much iodine is *more than* 2 × 30 seconds.

Example 9.2

The table of observations shown below is taken from a laboratory investigation into the rate of decomposition of hydrogen peroxide using a manganese(IV) oxide catalyst. For each set of initial conditions, the time taken to collect a fixed volume of gas was noted. In each experiment 50 cm^3 of aqueous hydrogen peroxide and 1 g of catalyst were used.

Experiment	Concentration of aqueous hydrogen peroxide in mol/dm^3	Initial temperature in °C	Form of the catalyst	Time, in seconds, to collect a fixed volume of gas
A	0.1	20	powder	200
B	0.1	30	powder	115
C	0.1	40	powder	60
D	0.2	20	powder	100
E	0.3	20	powder	65
F	0.4	20	pellets	130
G	0.4	20	powder	50

(a) Draw a labelled diagram of an apparatus which could be used to perform this experiment.   **(3 marks)**
See Fig. 9.3.

(b) Write an equation for the reaction which takes place when hydrogen peroxide decomposes, and describe one simple test to identify the gas evolved.   **(3 marks)**

$2H_2O_2(aq) \rightarrow 2H_2O(l) + O_2(g)$
The gas is oxygen and it relights a glowing splint.

(c) Which experimental results should be used to draw a graph to show the relationship between concentration of reactant and rate of reaction? Explain briefly the reasons for the choice. **(DO NOT DRAW A GRAPH.)**   **(4 marks)**

*Results **A**, **D**, **E** and **G** should be used because the initial temperature and the form of the catalyst remain the same and only the concentration of the aqueous hydrogen peroxide and the rate of reaction vary.*

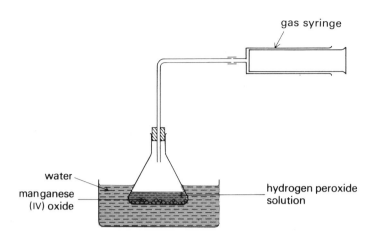

gas syringe

water

manganese (IV) oxide

hydrogen peroxide solution

Fig. 9.3

(d) (i) State and explain what happens to the rate of decomposition of the hydrogen peroxide when the concentration of its solution is increased.
  (ii) Explain why the use of catalyst pellets instead of powder alters the rate of reaction. (See experiments **F** and **G**.)
  (iii) State how one other factor investigated in this experiment affects the rate of reaction. **(7 marks)**

(i) *As the concentration of the solution increases, the time taken to collect a fixed volume of gas decreases: this means that the rate of reaction is increasing. The rate is faster because there are more hydrogen peroxide molecules per unit volume to collide with the catalyst and decompose.*

(ii) *If the 1 g of catalyst is in the form of pellets, it will have less surface area on which reaction can take place than if it were ground up into powder. Therefore the reaction is slower when the catalyst is pelleted.*

(iii) *As the temperature rises the reaction speeds up. This is because the molecules travel faster and collide more frequently and more vigorously with one another. The number of collisions per second which are sufficiently energetic to cause reaction is thereby increased, so the reaction speeds up.*

(e) Define the term *catalyst* and describe briefly the essential steps of an experiment to prove that manganese(IV) oxide acts as a catalyst in this reaction. **(7 marks)**
(AEB, 1981)

The definition of a catalyst is given in Section 9.1.
To show that manganese(IV) oxide acts as a catalyst in this experiment, the reaction should first be tried without the manganese(IV) oxide. The rate will be found to be very slow indeed — in fact it will probably appear that there is no reaction at all. The experiment should then be tried again, using a known mass of manganese(IV) oxide. The reaction will be quicker and, if the manganese(IV) oxide is filtered off at the end, washed, dried and reweighed, it will be found that its mass has not changed. Thus it will have satisfied the definition of a catalyst.

Example 9.3

The carbon dioxide produced by adding dilute hydrochloric acid to *excess* marble chips of uniform size was collected in a gas syringe and its volume read at regular intervals.

The experiment was repeated twice more at the same temperature and the results plotted as shown in Fig. 9.4.

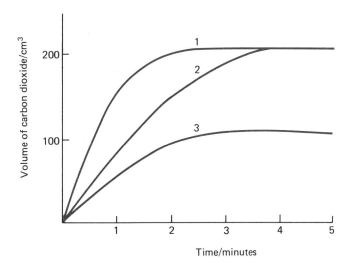

Fig. 9.4

The graph shows that:

A The number of moles of acid used in experiment 1 was greater than that used in 2.
B The number of moles of acid used in experiment 2 was less than that used in 1.
C The acid used in experiment 1 was more concentrated than that used in 2.
D The number of moles of acid used in experiment 3 was the same as that used in 1.

The answer is **C**.

The marble chips are in excess, therefore all of the acid is used up in each experiment. **A** and **B** are incorrect — the same volume of carbon dioxide was given off in 1 and 2, so the same number of moles of acid must have been used. **C** is correct — the reaction in 1 is faster, so the acid must be more concentrated than in 2. **D** is incorrect — less carbon dioxide is given, so fewer moles of acid were used.

Example 9.4

The formation of carbon monoxide and hydrogen from methane and steam at 750°C can be represented by the equation:

$$CH_4(g) + H_2O(g) \rightleftharpoons CO(g) + 3H_2(g); \quad \Delta H = -206 \text{ kJ}$$

(a) What mass of methane would have to react in order to produce 206 kJ of energy?

(1 mark)

1 mole of methane is needed, i.e. (12 + 4) = 16 g.

(b) What will happen to the rate of reaction if the pressure is increased? Explain. **(2 marks)**
The rate of reaction will become greater because an increase in pressure will increase the concentration of a gas.

9.5 Self-test Questions

Question 9.1

The reaction between 10 g of calcium carbonate and 10 cm^3 of hydrochloric acid was studied at 10°C. The experiment was then repeated at 15°C and 20°C. The results were plotted graphically and are represented by the figure. (AEB, 1983)

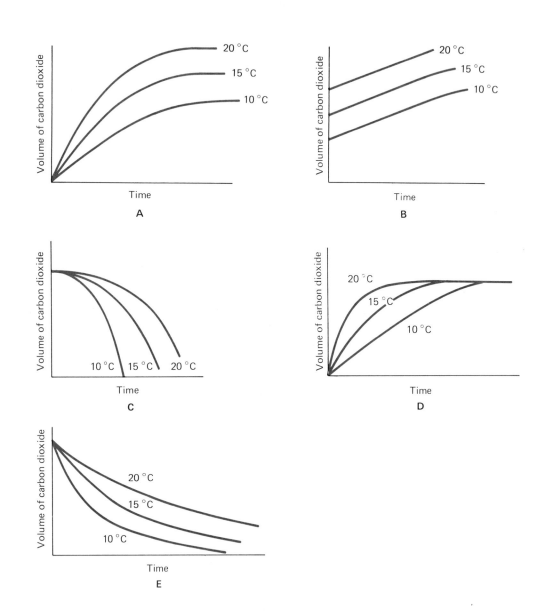

Question 9.2

The rate of production of hydrogen by the reaction of magnesium with 0.5 M hydrochloric acid (containing 0.5 mole of HCl per dm^3) was investigated using the apparatus shown.

The following results were obtained with a laboratory temperature of 15°C and 1 atmosphere pressure, using 0.0669 g of magnesium ribbon.

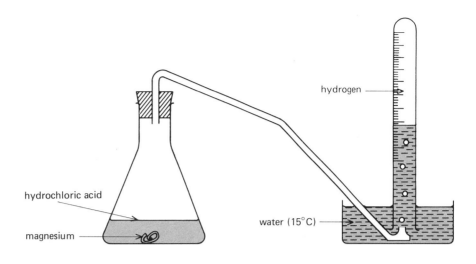

Time in minutes	0	$\frac{1}{2}$	1	$1\frac{1}{2}$	2	$2\frac{1}{2}$	3	$3\frac{1}{2}$	4	$4\frac{1}{2}$	5	$5\frac{1}{2}$
Total volume of H_2 in cm^3 (at 15°C and 1 atm.)	0	9.5	18.0	26.5	33.5	41.4	48.0	54.0	59.0	62.0	63.0	63.0

(a) (i) On squared paper, plot a graph of volume of hydrogen (vertical axis) against time (horizontal axis). **(4 marks)**

(ii) On the same graph paper, sketch and label the curve you would have expected if the hydrochloric acid had been heated to 25°C before the experiment. **(2 marks)**

(iii) What is the total volume of hydrogen obtained from 0.0669 g of magnesium at 15°C according to this experiment? **(2 marks)**

(iv) Calculate the volume that would have been obtained at this temperature from 1 mole of magnesium. **(4 marks)**

(v) How does your answer compare with the 24 dm^3 normally accepted for the molar volume of a gas? Give one possible reason for the discrepancy you have found. **(2 marks)**

(b) When a similar experiment was conducted with pieces of calcium carbonate instead of magnesium, a value of 16.8 dm^3 for the molar volume of carbon dioxide was found.

(i) Why do you think the value was so low? **(2 marks)**

(ii) State and explain two ways in which the rate of reaction of calcium carbonate with hydrochloric acid could be increased, other than by heating. **(4 marks)**

9.6 Answers to Self-test Questions

9.1 The answer is **A**.

Increasing the temperature increases the rate of reaction. The number of moles of carbon dioxide is the same in each case but the volume is greater at higher temperatures.

9.2 (a) (i) and (ii), see graph.

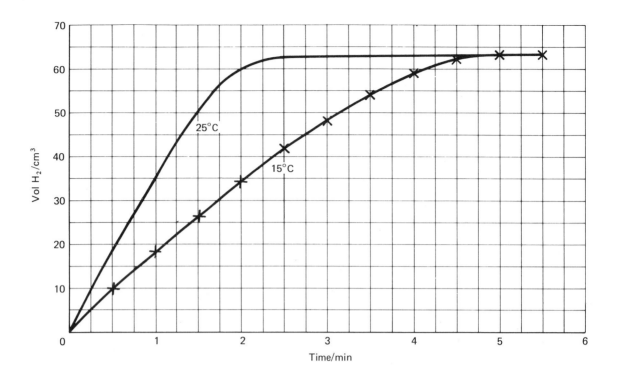

The reaction would be faster at 25°C because the particles in the acid would be speeded up and would collide more frequently and more violently with the magnesium. However, the final volume of hydrogen would be the same as before because the same amounts of magnesium and hydrochloric acid are used and the hydrogen is collected at the same temperature.

     (iii) 63.0 cm^3.

     (iv) Moles of Mg used = mass/molar mass (see Section 5.1)
                             = 0.0669/24

         0.0669/24 mol of Mg gives 63.0 cm^3 of H_2.

         $\therefore$ 1 mol of Mg gives 63.0 ÷ 0.0669/24 = 22 600 cm^3 of H_2

     (v) The value is lower than expected. Possibly the magnesium had an oxide coating so that the actual mass of *metal* was less than 0.0669 g.

(b) (i) Carbon dioxide is slightly soluble in water and could dissolve in both the dilute acid and the water in the trough.

     (ii) The rate could be increased by using more concentrated acid so that more H^+(aq) ions would be colliding with the surface of the calcium carbonate.

          If the calcium carbonate were broken up into smaller pieces, there would be a greater surface area in contact with the acid and the reaction would be speeded up.

10 The Atmosphere and Combustion

10.1 Introduction

Air consists of a mixture of gases, the percentages being as shown in Fig. 10.1.

Water vapour will also be present (cobalt(II) chloride paper turns pink in the air), the exact proportion depending on conditions.

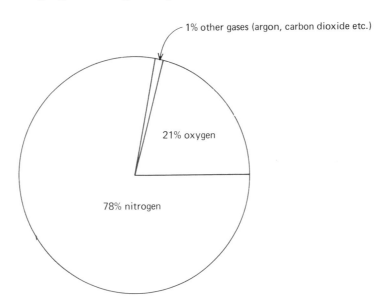

1% other gases (argon, carbon dioxide etc.)

21% oxygen

78% nitrogen

Fig. 10.1

Air in built-up areas will be polluted by fumes from cars, fires, factories, etc. (see Example 10.7). Substances present may include dust, soot, lead compounds from car-exhaust gases, carbon monoxide (see Section 14.6), unburnt hydrocarbons from petrol and other fuels, nitrogen dioxide (see Section 15.2) and sulphur dioxide (see Section 16.5). Most of these are corrosive and/or poisonous. Finally, there are usually bacteria and viruses which may or may not be harmful. Pollution must be controlled, and this can be achieved by means of, for example, smokeless zones, lead-free petrol and control of exhaust emissions.

N.B. Hydrogen is not generally found in the atmosphere.

10.2 Separating the Gases in the Air

Since air is a mixture of gases, it can be separated into its components by physical means: fractional distillation of liquid air. The air is cooled from room temperature to about $-200°C$, ice being removed at $0°C$ and solid carbon dioxide at $-78°C$. Fractional distillation then takes place to yield nitrogen (b.p. $-196°C$), argon (b.p. $-186°C$) and finally oxygen (b.p. $-183°C$).

10.3   Processes which use up Oxygen

There are three main ways in which oxygen is used up: burning, corrosion and respiration. If a piece of iron is allowed to rust or a substance is allowed to burn in a closed container, then the percentage of oxygen in the air can be found, as Examples 10.2 and 10.6 show.

(a) Burning (Combustion)

When a substance burns in air, it combines with oxygen (i.e. oxidation occurs). Combustion continues until all of the substance or all of the oxygen is used up (see Example 10.5).

(b) Rusting

Iron will only rust if both air and water are present; dissolved salt is found to speed up the process (see Example 10.6). The gas in the air which is responsible for rusting is oxygen. Rusting, like combustion, is an example of oxidation, but in this case, the temperature is much lower and the reaction occurs much more slowly.

Many metals tarnish in air but the coating formed on the surface often prevents further corrosion, e.g. zinc, copper and aluminium. However, rust does not prevent air and water from reaching the iron underneath.

(c) Respiration

When we and other animals breathe, the air we breathe out is warmer and contains more water vapour and carbon dioxide than it did when we inhaled it. We are 'burning' food inside us to provide energy for movement and warmth (see Example 10.5). The overall change may be summed up as

sugar + oxygen → carbon dioxide + water + energy

10.4   Photosynthesis

Combustion, rusting and respiration remove oxygen from the atmosphere. Photosynthesis replaces the oxygen.

Photosynthesis is the process by which green plants synthesise carbohydrates from carbon dioxide and water, using sunlight as the source of energy and chlorophyll as a catalyst, oxygen being liberated into the air.

The overall change may be written:

carbon dioxide + water + energy → sugar + oxygen

which is the reverse of the equation summarising respiration.

10.5 Worked Examples

Example 10.1

If air is bubbled through pure water (pH 7), the pH gradually changes to 5.7. The gas in air which is responsible for this change is

A carbon dioxide
B nitrogen
C hydrogen
D argon
E oxygen.   (L)

The answer is **A**.

Air contains carbon dioxide and this is an acidic oxide, dissolving in water to give carbonic acid.

Example 10.2

In order to find the proportion by volume of one of the main constituents of air, a sample of air was passed through two wash bottles, the first containing aqueous sodium hydroxide and the second containing concentrated sulphuric acid, and was then collected in a gas syringe.

(a) Suggest a reason for passing the air through

    (i) aqueous sodium hydroxide

To remove carbon dioxide.

    (ii) concentrated sulphuric acid.   **(2 marks)**

To dry the gas.

(b) The volume of gas collected in the syringe was 80 cm^3. This was passed several times over hot copper powder until no further contraction of volume took place. After cooling to the original temperature the volume was found to be reduced to 63.2 cm^3.

    (i) How would the copper change in appearance?

It would turn black.

    (ii) Which gas had been removed by the copper?

Oxygen.

    (iii) Calculate the volume of this gas present in the sample of air.

$80 - 63.2 \ cm^3 = 16.8 \ cm^3$.

    (iv) Calculate the percentage of this gas present in the sample of air.   **(4 marks)**

$$\% \ Oxygen = \frac{Volume \ of \ oxygen}{Original \ volume \ of \ air} \times 100$$

$$\% \ Oxygen = \frac{16.8}{80} \times 100 = 21$$

(c) (i) What is the main gas remaining in the syringe?

Nitrogen.

    (ii) There will be small amounts of other gases remaining in the syringe. Name one of these gases.   **(2 marks)**

Argon.

In addition to nitrogen there are traces of the noble gases, e.g. argon, neon, helium.

    (iii) Give one use of argon.   **(1 mark)**

In electric light bulbs.

(d) How would you treat the solid remaining to obtain copper for re-use? **(1 mark)**

(L)

The solid remaining is copper(II) oxide. It needs to be reduced to copper, e.g. by heating it in a stream of hydrogen.

Example 10.3

·Explain why magnesium gains in mass but coal loses mass when each is burned in air.

(2 marks)

Both substances combine with oxygen when they burn but the magnesium forms a solid product which can be weighed. The coal forms some gaseous products which go into the atmosphere and are not weighed.

Example 10.4

Fig. 10.2

(a) What is the remaining part of the fire triangle (Fig. 10.2)? **(1 mark)**

Air or oxygen.

For combustion to occur, the three components of the fire triangle are needed. The three requirements of fire fighting are exclusion of air, removal of fuel and temperature reduction.

(b) A small child has been playing with matches and his clothes have caught fire. Explain how you would put out the flames. **(2 marks)**

Air must be excluded as quickly as possible so that burning will cease. Therefore the child must be wrapped in something like a rug or a blanket or turned over so that the flames are between his body and the floor.

Example 10.5

The equation for the complete oxidation of glucose is

$$C_6H_{12}O_6 + 6O_2 \rightarrow 6CO_2 + 6H_2O$$

This is an exothermic process.

(a) Use this process, or any other, to explain the difference between *combustion* and *respiration*. **(5 marks)**

When combustion takes place, the glucose burns in air and combines with oxygen. When we breathe, the same reaction occurs. The difference is that in respiration, the reaction occurs more slowly and at a lower temperature than in combustion.

(b) (i) What is meant by the term *exothermic*? **(2 marks)**
An exothermic reaction is one which gives out heat (see page 72).

(ii) State two other processes, physical or chemical, that are exothermic. **(2 marks)**
Exothermic processes include neutralisation of an acid by an alkali,
e.g. HCl*(aq)* + NaOH*(aq)* → NaCl*(aq)* + H$_2$O*(l)*
and passing an electric current through a wire, e.g. in an electric fire.

(c) (i) Give the general name for the process by which glucose is converted into ethanol by yeast. **(1 mark)**
Fermentation (see Section 19.5).

(ii) Write an equation for this process. **(2 marks)**
(OLE)

C$_6$H$_{12}$O$_6$*(aq)* → 2C$_2$H$_5$OH*(aq)* + 2CO$_2$*(g)*

Example 10.6

(a) Name the two substances which are essential if iron is to rust. **(2 marks)**
(i) *Water.* (ii) *Oxygen.*

(b) Name one other substance which will accelerate the rusting process. **(1 mark)**
Salt

(c) Why is the use of this substance of particular importance to motorists? **(1 mark)**
Salt is put on the roads in winter to melt the ice: an ice/salt mixture has a lower melting point than pure ice. The salt must be washed off to prevent rapid rusting of the car.

(d) Give three methods which can be used to prevent rusting. **(3 marks)**
(i) *Painting.*
(ii) *Coating with grease or oil.*
(iii) *Coating with metal, e.g. tin plating or galvanising.*

In order to prevent rusting, air and water must be kept away from the iron.

(e) Draw labelled diagrams to show three experiments you could do to show that the substances mentioned in (a) (i) and (ii) are needed for iron nails to rust, and that if either of the substances is absent, no rusting occurs. **(3 marks)**
See Fig. 10.3.

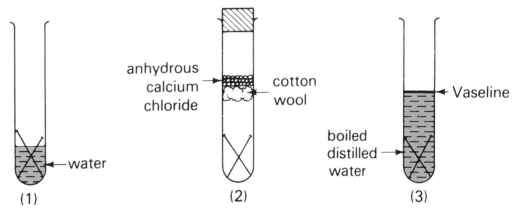

anhydrous calcium chloride

cotton wool

Vaseline

boiled distilled water

water

(1)   (2)   (3)

Fig. 10.3

Experiment 1
(a) (i) and (ii) present
-- rusting occurs

Experiment 2
(a) (i) absent
— no rusting

Experiment 3
(a) (ii) absent
— no rusting

Some moist iron wool was placed in a test tube and the tube was inverted and placed in a beaker of water (Fig. 10.4). The apparatus was inspected each day for one week. The iron rusted and the level of the water in the tube rose during the first few days. After this, no further change took place, even though some air remained and not all of the iron was rusty. The air left in the tube was found to put out a burning splint.

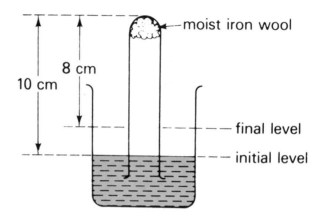

Fig. 10.4

(f) What fraction of the air, approximately, has been used up? **(1 mark)**

The water has moved 2 cm up the tube.

$$\% \text{ Air used} = \frac{2}{10} \times 100 = 20\%$$

(g) Does this experiment show that the air is made up of (i) one, (ii) two, (iii) at least two, gases? Explain your answer. **(2 marks)**

At least two gases. One gas has been used up but we have no means of knowing in this experiment how many gases remain in the tube.

(h) Which gas has been used up during rusting? How can you tell? **(2 marks)**

Oxygen. Oxygen is needed for a splint to burn; since the splint went out, the air must no longer contain oxygen.

(i) What would be the effect on the level of the water if a larger piece of iron wool were used? How did you reach your conclusion? **(2 marks)**

No effect. Not all the iron had rusted and so a larger piece of iron would make no further change.

Example 10.7

In America cars are fitted with filters in their exhaust systems so that nitrogen monoxide and carbon monoxide are converted into relatively harmless substances.

$$2CO(g) + 2NO(g) \rightarrow 2CO_2(g) + N_2(g)$$

(a) How is nitrogen monoxide formed in the engine? **(1 mark)**

Nitrogen and oxygen combine at the high temperature inside the engine to give nitrogen monoxide.

(b) What happens if nitrogen monoxide is released into the atmosphere? **(2 marks)**

It combines with oxygen to produce nitrogen dioxide. This is a respiratory irritant and also dissolves in water to produce acid rain.

(c) Where does the carbon monoxide come from? **(1 mark)**

It is formed by the partial combustion of petrol.

(d) Why is carbon monoxide considered to be harmful? **(2 marks)**

It is poisonous but it is also odourless.

(e) Carbon dioxide could also be considered to be a pollutant. Explain why. **(2 marks)**

The carbon dioxide layer in the atmosphere results in the 'greenhouse effect'. The sun's rays pass through the carbon dioxide layer. The heat carried by these rays is absorbed by objects on the ground which then reradiate it at a longer wavelength. This new radiation cannot pass back through the carbon dioxide layer and thus the atmosphere slowly heats up.

(f) Name one other pollutant that could be found in the exhaust gases from British cars.
(1 mark)

Lead compounds from petrol additives.

(g) How could this pollutant be eliminated? **(1 mark)**

The use of lead-free petrol.

10.6 Self-test Questions

Questions 10.1 and 10.2 concern the following account.

A strip of copper metal 5 cm by 1 cm was folded over to make a square 1 cm by 1 cm and pressed down tightly. It was heated in a bunsen burner and the outside went black. It was allowed to cool down and then unfolded. The inside was still copper coloured.

Question 10.1

The outside was black and the inside copper coloured because:
A the inside was not hot enough
B air reached the inside but not the outside
C air reached the outside but not the inside
D oxygen reached the outside and nitrogen reached the inside
E the air removed something from the outside of the copper.

Question 10.2

This experiment tells you that the percentage of oxygen in the air is about:
A 20%
B 40%
C 60%
D 80%
E no conclusion can be drawn.

Question 10.3

(a) Which two substances are necessary for iron to rust? **(2 marks)**
(b) Why does tin plating prevent iron from rusting? **(1 mark)**

(c) If the tin surface is scratched then the iron will rust. Why does this not happen with zinc plating (galvanising)? **(2 marks)**

Question 10.4

Which one of the following would not produce carbon dioxide?
A heating carbon in air
B reacting limewater with air
C burning starch in air
D burning a candle in air
E breathing.

Question 10.5

Complete the following sentences.
(a) The most abundant gas in the air is _____ .
(b) Liquid air is separated into its components by _____ _____ .
(c) _____ is the process by which green plants synthesise carbohydrates from carbon dioxide and water, using sunlight as the source of energy, and chlorophyll as a catalyst.
(d) A pollutant in the air which leads to the production of acid rain is _____ _____ .
(e) The incomplete combustion of diesel fuel produces a poisonous gas called _____ _____ . **(5 marks)**

10.7 Answers to Self-test Questions

10.1 C.

10.2 E.

Do not be trapped into thinking that the answer is **A**. The percentage of oxygen in the air *is* approximately 20% but this experiment does not tell you this.

10.3 (a) Water and oxygen.
  (b) It covers the iron and stops the air and water reaching the iron.
  (c) Zinc is above iron in the reactivity series and so will tend to pass into solution rather than the iron which remains intact. Tin is below iron and so will not corrode in preference to the iron.

10.4 B.

Limewater goes milky when carbon dioxide is passed through it — it absorbs carbon dioxide.

10.5 (a) Nitrogen.
  (b) Fractional distillation.
  (c) Photosynthesis.
  (d) Sulphur dioxide.
  (e) Carbon monoxide.

11 Water and Hydrogen

11.1 Introduction

Pure water, formula H_2O, is a colourless, odourless, tasteless liquid which freezes at $0°C$ and boils at $100°C$ at 1 atm pressure (see Question 11.1). Water is unusual in that it expands on freezing whereas most liquids contract. Thus ice is less dense than water and forms on the surface rather than at the bottom of the liquid.

Water reacts with reactive metals, non-metals and various compounds (see Example 11.4), e.g.

$$2Na(s) + 2H_2O(l) \rightarrow 2NaOH(aq) + H_2(g) \qquad \text{(see Section 13.2)}$$
$$SO_2(aq) + H_2O(l) \rightleftharpoons H_2SO_3(aq) \qquad \text{(see Section 16.5)}$$

(a) Tests for Water (see Example 11.1)

White anhydrous copper(II) sulphate turns blue and blue cobalt(II) chloride turns pink when water is added.

$$CuSO_4(s) + 5H_2O(l) \rightleftharpoons CuSO_4.5H_2O(s)$$
$$\text{white} \qquad\qquad\qquad\quad \text{blue}$$
$$CoCl_2(s) + 6H_2O(l) \rightleftharpoons CoCl_2.6H_2O(s)$$
$$\text{blue} \qquad\qquad\qquad\quad \text{pink}$$

These two reactions simply prove that a liquid *contains* water. To prove that a liquid is *pure* water, a boiling point determination must be carried out as well (see Question 11.1).

(b) Uses of Water

Water is essential for all life: over two-thirds of your own body mass is water. Water is extensively used in industry for heating and cooling, for producing steam to drive turbines and as a solvent. It is a raw material in the manufacture of ammonia (see Example 18.6) and ethanol (see Section 19.5). In addition, vast amounts are used in the home.

11.2 Water as a Solvent

A **solution** consists of a **solute** dissolved in a **solvent**; if the solvent is water, then an aqueous solution is obtained.

A **saturated solution** is one that contains as much solute as can be dissolved at the temperature concerned, in the presence of undissolved solute.

The **solubility** of a solute in a solvent at a particular temperature is the mass of solute required to saturate 100 g of solvent at that temperature (see Example 11.3).

For most solids, the solubility in a given solvent increases as the temperature is raised but the reverse is true for gases: the solubility decreases with rise in temperature (see Question 11.3). The gases which are most soluble in water are those that react with it in some way, e.g. ammonia, hydrogen chloride.

A graph showing the variation of solubility with temperature is known as a solubility curve (see Example 11.3).

Many crystals contain water chemically combined within them. This water is referred to as water of crystallisation, and salts containing it are called hydrated salts. Water of crystallisation can often be driven off by gentle heating, and when this is done it is found that there is a definite number of water molecules associated with every formula unit of the compound concerned.

Water of crystallisation is that definite quantity of water with which some substances are associated on crystallising from an aqueous solution.

11.3 Water Supplies

Our water supplies come from rivers, lakes and underground sources. These are replenished by the rain by means of the water cycle (Fig. 11.1).

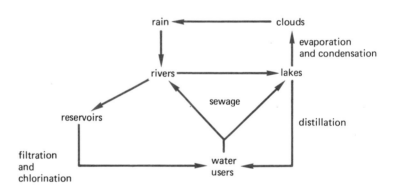

Fig. 11.1

When water evaporates from a sea or lake, the vapour rises and condenses to form many very small droplets (a cloud). If these droplets come together, they fall as rain which eventually returns to a sea or lake so that the cycle can begin again.

In recent years, water pollution has become a serious problem. Water contains dissolved oxygen which is used by fish and other aquatic creatures and also by bacteria which feed on animal and vegetable remains and keep our rivers healthy. Too much untreated sewage causes the bacteria to multiply and they may use up so much oxygen that there is insufficient left for fish to survive. Fertilisers, insecticides, detergents and industrial waste all help to upset the delicate balance of life in the water (see Example 11.5).

Our drinking water must contain none of the pollutants mentioned above. The water is filtered and chlorine added to kill bacteria. In some areas, a trace of sodium fluoride is added to reduce dental decay. Tap water contains other dissolved solids, particularly calcium and magnesium salts (see Section 13.3).

11.4 Hydrogen, H_2

Hydrogen is generally prepared in the laboratory by the action of dilute sulphuric acid on zinc in the apparatus shown in Example 11.6.

$$Zn(s) + H_2SO_4(aq) \rightarrow ZnSO_4(aq) + H_2(g)$$

It is also a product in the electrolysis of acidified water (see Section 7.2) and of the reaction of metals with water and/or steam (see Section 13.3 and Example 11.4).

(a) Test for Hydrogen

A mixture of hydrogen with air or oxygen explodes when a flame is applied.

$$2H_2(g) + O_2(g) \rightarrow 2H_2O(g)$$

(b) Properties of Hydrogen

1. Hydrogen is a colourless, odourless gas.
2. It is almost insoluble in water.
3. It is less dense than air.
4. Pure hydrogen burns quietly in air with a faint, blue flame to give water (see Question 11.1).
5. On heating, hydrogen reduces the oxides of metals low in the reactivity series (see Example 11.6), e.g.

$$CuO(s) + H_2(g) \rightarrow Cu(s) + H_2O(g)$$

6. Hydrogen reacts with some metals and non-metals (see Example 11.7), e.g.

$$N_2(g) + 3H_2(g) \rightleftharpoons 2NH_3(g) \qquad \text{(see Example 18.6)}$$

(c) Uses of Hydrogen (see Example 11.7)

1. In the large-scale syntheses of ammonia and hydrogen chloride (see Example 18.6 and Section 17.1).
2. In the hardening of oils to make margarine and cooking fats (see Example 18.4).
3. As a fuel. There are no pollutant products but there is always the risk of explosion.

(d) Manufacture of Hydrogen

Hydrogen is produced on a large scale:

1. in the cracking of oils (see Section 19.4);
2. in the electrolysis of brine (see Example 18.4); and
3. by the action of steam on hydrocarbons such as methane (North Sea Gas).

11.5 Worked Examples

Example 11.1

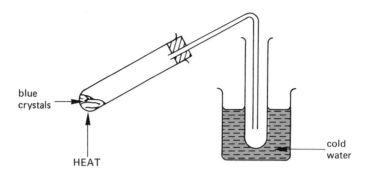

Fig. 11.2

The apparatus is set up as shown in Fig. 11.2, and the left-hand tube is gently heated.

(a) Why is the right-hand tube placed in a beaker of water? **(2 marks)**

To condense the vapour that is given off when the crystals are heated.

(b) What colour changes are seen when the crystals are heated? **(2 marks)**

The crystals change from blue to white.

(c) Draw a diagram (Fig. 11.3) to show how you can measure the boiling point of the product. **(2 marks)**

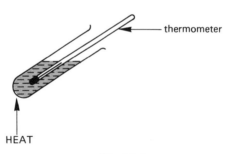

Fig. 11.3

(d) The boiling point of the liquid is 100°C. Suggest what the liquid might be. **(1 mark)**

Water.

(e) Name the original crystals. **(1 mark)**

Copper(II) sulphate crystals.

(f) What would happen if a piece of cobalt(II) chloride paper were placed in the liquid?

(2 marks)

The cobalt(II) chloride paper would change from blue to pink.

This test shows the presence of water in a liquid.

Example 11.2

Which one of the following is insoluble in water?

A zinc nitrate

B zinc carbonate

C sodium carbonate

D sodium sulphate
E potassium chloride
The answer is **B.**

All nitrates and all sodium, potassium and ammonium salts are soluble in water.

Example 11.3

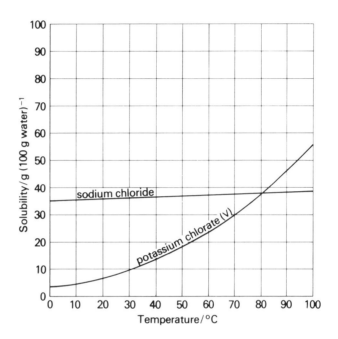

Fig. 11.4

Look at the graph shown in Fig. 11.4, and then answer the following.

(a) Which compound is the more soluble in boiling water?   **(1 mark)**

Potassium chlorate(V).

(b) At what temperature are sodium chloride and potassium chlorate(V) equally soluble in
water?   **(1 mark)**

81°C.

(c) A mixture of 30 g sodium chloride, 30 g potassium chlorate(V), and 100 g of water, is
stirred for a long time at 100°C. Will all the solid dissolve? If the mixture is now cooled
to 20°C, what will happen?   **(4 marks)**

Yes.

At 100°C, the solubilities of both potassium chlorate(V) and sodium chloride
are above 30 g (100 g water)$^{-1}$ and therefore all the solid dissolves.

---

*At 20°C, the solubility of sodium chloride is 36 g (100 g water)$^{-1}$ whilst that of
potassium chlorate(V) is 7 g (100 g water)$^{-1}$. This means that all the sodium
chloride will remain in solution but 23 g of potassium chlorate(V) will crystallise
out.*

Example 11.4

Steam is passed over heated magnesium in the apparatus shown in Fig. 11.5.

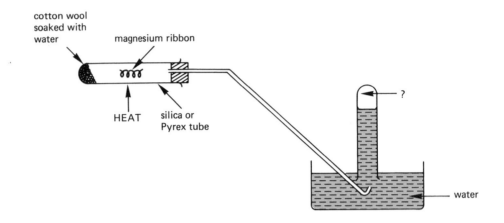

Fig. 11.5

(a) Name the gas produced. **(1 mark)**
Hydrogen.

(b) How could you test for this gas? **(2 marks)**
If a flame is applied to a tube of the gas, a 'pop' is heard.

(c) What is the other product of the reaction between magnesium and steam? **(1 mark)**
Magnesium oxide.

(d) Write a word equation for the reaction. **(2 marks)**
Magnesium + steam → magnesium oxide + hydrogen

(e) Write a symbol equation for the reaction. **(2 marks)**
$Mg(s) + H_2O(g) \rightarrow MgO(s) + H_2(g)$

(f) Would you pass steam over sodium in a similar way? Explain your answer. **(2 marks)**
No, since the reaction is too violent.

Example 11.5

Excess nitrates or excess heat in a river can be regarded as pollution. In each case, describe **one** possible source of this pollution and the effect on the river and its environment.
(a) Various nitrates.
Source: *Fertilisers, e.g. ammonium nitrate or sodium nitrate. Fertilisers are added to soil, but heavy rain has the effect of washing them into rivers.* **(1 mark)**
Effect: *The addition of fertilisers to a river has the effect of upsetting the ecological balance of the river by promoting the too rapid growth of lower plant life. For example, vast amounts of algae could grow on the surface of the river, thus reducing the amount of light to plants on the bottom of the river. These plants decay, no longer generate oxygen and thus the animals in the river die.* **(2 marks)**

(b) Excess heat.
Source: *Industrial effluent. Many industrial processes use water as a coolant. If this hot water is discharged into a river, then the temperature increases.* **(1 mark)**
Effect: *As long as the temperature increase is moderate, the effect is generally beneficial since the growth of all forms of aquatic life is promoted.* **(2 marks)**

(c) (i) Name one other pollutant of river water. **(1 mark)**
Phosphates.

(ii) What is the source and what is the effect of this pollutant? **(3 marks)**
Source: *Washing powders.*

Effect: *Like nitrates, phosphates are plant nutrients and encourage the growth of algae.*

(d) (i) How is reservoir water filtered? **(1 mark)**
The water is passed through gravel or sand.

(ii) Why is this process carried out? **(2 marks)**
To remove solids and to enable oxygen from the air to kill some bacteria.

(iii) Explain why the water is treated with chlorine. **(2 marks)**
Chlorine kills bacteria.

Household tap water contains neither dirt nor harmful bacteria.

Example 11.6

(a) How would you prepare hydrogen in the laboratory by
(i) electrolysis, **(2 marks)**
Hydrogen may be obtained by the electrolysis of dilute sulphuric acid using platinum or carbon electrodes (see Section 7.2).
At cathode $4e^- + 4H^+(aq) \rightarrow 4H(g)$
$$\downarrow$$
$$2H_2(g)$$

(ii) some other method?
Write an equation for the reaction in which hydrogen is produced and give the reagents and necessary conditions. **(3 marks)**
Hydrogen is generally prepared by the action of dilute sulphuric acid on zinc in the apparatus shown in Fig. 11.6. The addition of a little copper(II) sulphate solution will speed up the reaction.

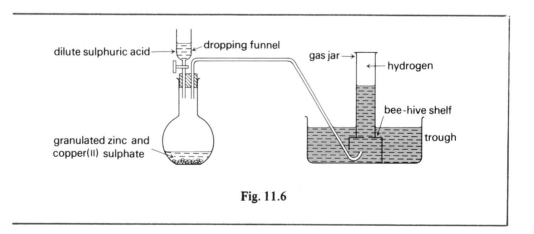

Fig. 11.6

$Zn(s) + H_2SO_4(aq) \rightarrow ZnSO_4(aq) + H_2(g)$

(b) A current of dry hydrogen was passed over 4.78 g of heated lead oxide, the excess hydrogen being burnt in the air. 4.14 g of lead was obtained.
(i) Calculate the empirical formula of the oxide. **(3 marks)**

$A_r(O) = 16$; $A_r(Pb) = 207$

Mass of oxygen combined with 4.14 g of lead = 4.78 − 4.14 = 0.64 g

ratio of g   Pb : O

ratio of g   4.14 : 0.64

ratio of mol $\dfrac{4.14}{207}$: $\dfrac{0.64}{16}$

= 0.02 : 0.04

= 1   : 2

Empirical formula is PbO_2 *(see Section 5.1)*

    (ii) Name one other oxide which can be reduced to the metal by hydrogen.   **(1 mark)**
Copper(II) oxide

    (iii) Name one oxide which cannot be reduced to the metal by hydrogen.   **(1 mark)**
Sodium oxide.

Hydrogen will reduce only the oxides of those metals which are low in the re-activity series (see Section 13.1).

    (c) (i) Write an equation for the reaction between hydrogen and ethene. Name the product and give the necessary conditions.   **(2 marks)**

$H_2(g) + C_2H_4(g) \rightarrow C_2H_6(g)$

               ethane

Hydrogen adds on to ethene in the presence of a finely divided nickel catalyst at about 200°C.

    (ii) In the margarine industry, this type of reaction is used to convert animal and vegetable oils into solid fats. What does this tell you about the bonding in natural oil molecules?   **(1 mark)**

This tells us that natural oil molecules must contain C=C bonds, i.e. they are unsaturated.

    (d) There are vast reserves of hydrogen in sea water, and its use as a fuel carries no pollution risk. Suggest one reason in each case why

    (i) we have not obtained hydrogen on a large scale from water.   **(1 mark)**

Hydrogen can be obtained from sea water by electrolysis but the cost of electricity makes this method too expensive.

    (ii) its use as a fuel carries no pollution risk.   **(1 mark)**

When hydrogen is burnt, the only product is water, which is not a pollutant.

$2H_2(g) + O_2(g) \rightarrow 2H_2O(g)$

Example 11.7

    (a) (i) Describe, with the aid of a balanced equation, a reduction reaction which does not involve the use of hydrogen gas.   **(6 marks)**

If carbon monoxide is passed over heated black copper(II) oxide, red/brown copper is obtained, the excess carbon monoxide being burnt.

$CuO(s) + CO(g) \rightarrow Cu(s) + CO_2(g)$

    (ii) Explain why the reaction which you have described can be classified as a reduction process.   **(2 marks)**

The copper(II) oxide has lost oxygen and has therefore been reduced.

    (b) (i) Write an equation for the reaction between hydrogen and chlorine.   **(2 marks)**

$H_2(g) + Cl_2(g) \rightarrow 2HCl(g)$

    (ii) What volume of product would be obtained by complete reaction of 12 dm³ of hydrogen, all gaseous volumes being measured at room temperature and pressure.   **(3 marks)**

mol $H_2(g)$: *mol* $HCl(g) = 1:2$
vol $H_2(g)$: *vol* $HCl(g) = 1:2$
Volume of product = 24 dm³ (see Section 6.1).

     (iii) What **mass** of chlorine would be reduced by this volume of hydrogen?   **(3 marks)**

dm^3 $H_2(g) = 12$
mol $H_2(g) = \frac{12}{24} = \frac{1}{2}$
mol $Cl_2 = \frac{1}{2}$
       $Cl_2(g) = \frac{1}{2} \times 71$ $(M_r(Cl_2(g)) = 71)$
           $= 35.5$

Mass of chlorine reduced by 12 dm³ of hydrogen = 35.5 g.

    (c) Write brief notes on **two** large-scale uses of hydrogen.   **(4 marks)**
                                                                                 (OLE)

Ammonia is made industrially by combining hydrogen with nitrogen at a temperature of 500°C and a pressure of 250 atm in the presence of an iron catalyst (see Section 18.5).
$N_2(g) + 3H_2(g) \rightleftharpoons 2NH_3(g)$
Hydrogen is also used to convert animal and vegetable oils (which contain carbon to carbon double bonds) into solid fats (see Example 11.6).

Example 11.8

    (a) Figure 11.7 shows the apparatus used to determine the composition of water by mass.

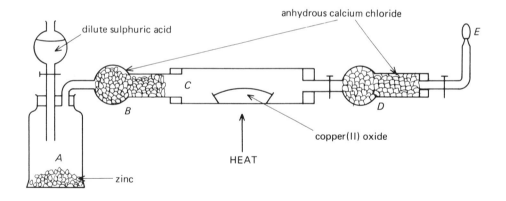

Fig. 11.7

     (i) Name the gas evolved in *A*.
Hydrogen.

     (ii) Write the equation for the reaction in *A*.
$Zn(s) + H_2SO_4(aq) \rightarrow ZnSO_4(aq) + H_2(g)$

     (iii) What is the purpose of *B*?
The anhydrous calcium chloride in B is used to dry the hydrogen. This ensures that the water absorbed in D has been produced in the reaction in C and was not present in the hydrogen initially.

(iv) Write the equation for the reaction in *C*.

$$CuO(s) + H_2(g) \rightarrow Cu(s) + H_2O(g)$$

(v) What is the purpose of *D*?

To absorb the water produced in C.

(vi) Why is it necessary to allow the reaction in *A* to continue for some considerable time before the excess gas is ignited at jet *E* and the copper oxide is heated?

Initially the apparatus is filled with air; air/hydrogen mixtures are explosive. Therefore air must be expelled from the apparatus before the hydrogen is ignited.

(vii) At the end of the experiment it was found that the increase in mass of *D* was 0.9 g, and the decrease of the copper(II) oxide was 0.8 g. What do these figures give for the composition of water by mass? **(10 marks)**

Mass of water produced   *= 0.9 g*
Mass of oxygen lost by copper(II) oxide = mass of oxygen in water = 0.8 g
Mass of hydrogen in water   *= mass of water − mass of oxygen*
                                               = 0.9 − 0.8 g
                                                 = 0.1 g
Composition of water by mass   *= 0.1 g hydrogen : 0.8 g oxygen*
                                                   = 1 : 8

(b) Draw a labelled diagram of the apparatus you would use to determine the composition of water by volume. **(5 marks)**

(SUJB)

See Fig. 7.2 and Section 7.2.

Either of these sets of apparatus illustrated in Fig. 7.2 is suitable. The electrolyte is sodium fluoride solution, and graduated test tubes are used to collect the gases. Pure water is a very poor conductor of electricity but its conductivity is improved by dissolving a substance such as sodium fluoride in it. The sodium and fluoride ions are not discharged and so do not affect the overall result. When this experiment is carried out, it is found that

volume of hydrogen : volume of oxygen = 2 : 1.

11.6   Self-test Questions

Question 11.1

A pupil set up the apparatus shown in the figure.

(a) Name the substance labelled A in the figure, and state **one** physical test which would help to identify it. **(3 marks)**

(b) Explain the significance of the tube filled with anhydrous calcium chloride granules. **(2 marks)**

(c) For each of the following substances, name an alternative reagent which could be used successfully in its place:

    (i) impure zinc; **(1 mark)**

    (ii) dilute sulphuric acid; **(1 mark)**

    (iii) anhydrous calcium chloride granules. **(1 mark)**

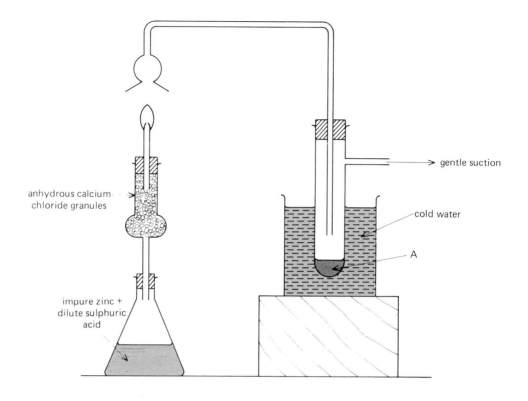

anhydrous calcium chloride granules

impure zinc + dilute sulphuric acid

gentle suction

cold water

A

(d) Write down a balanced equation to represent the reaction between the substances named in your answers to (c) (i) and (c) (ii) above. **(1 mark)**
(e) Write down an aspect of chemistry which the pupil could demonstrate with this equipment. **(2 marks)**

(OLE)

Question 11.2

Fill in the blanks in the following sentences:
(a) The element used to sterilise water in swimming pools is _____ .
(b) _____ is the element in drinking water which is good for the growth of healthy teeth.
(c) _____ is an element which can be extracted from sea water.
(d) _____ _____ is a solid used as a drying agent for gases.
(e) An ion which causes hardness in water (see Section 13.3) is _____ .
(f) When hydrochloric acid reacts with magnesium, the gas evolved is _____ .
(g) Pure water can be obtained from sea water by_____. **(1 mark each)**

Question 11.3

The solubility of nitrogen monoxide in water at room temperature is one volume to one volume. Given that 1 mol of nitrogen monoxide occupies 24.0 dm^3 at room temperature and pressure, how many moles of the gas will dissolve in 1 dm^3 of water under these conditions? How would an increase in temperature affect the solubility? **(3 marks)**

(SUJB)

Question 11.4

A desert survival kit contains a plastic sheet and a cup. A hole is dug in the ground and the sheet is stretched over the hole whilst a stone in the middle forms the sheet into a cone. In the heat of the sun, moisture from the ground collects on the underside of the sheet, runs down to the point of the cone and drips into the cup. The overall process that takes place is an example of

A distillation

B evaporation

C filtration

D solution

E solvent extraction.

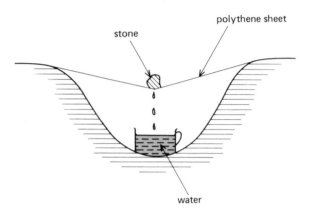

(AEB, 1981)

11.7 Answers to Self-test Questions

11.1 (a) Water. The boiling point of the liquid could be found — water boils at 100°C at 1 atm pressure.

(b) Anhydrous calcium chloride is used to dry the hydrogen. It is no use trying to prove that water is produced if water is present already.

(c) (i) iron;

(ii) dilute hydrochloric acid;

(iii) potassium hydroxide pellets or calcium oxide.

(d) $Fe(s) + 2HCl(aq) \rightarrow FeCl_2(aq) + H_2(g)$

(e) Water is obtained by burning hydrogen in air.

11.2 (a) Chlorine.

(b) Calcium.

(c) Either sodium or chlorine.

(d) Either anhydrous calcium chloride or calcium oxide.

(e) Calcium.

(f) Hydrogen.

(g) Distillation.

The apparatus used in Fig. 1.6 may be used. The flask contains impure water; pure water will distil over.

11.3 Nitrogen monoxide : Water

           1 volume : 1 volume

           1 dm^3 : 1 dm^3

Vol nitrogen monoxide   = 1 dm^3

Moles nitrogen monoxide = $\frac{1}{24}$

$\frac{1}{24}$ mol nitrogen monoxide will dissolve in 1 dm^3 water.

Generally, an increase in temperature decreases the solubility of gases.

11.4 A.

12 Acids, Bases and Salts

12.1 Introduction

Acidity is measured on the **pH scale**, a scale of numbers normally ranging from 0 to 14. A pH value of 7 is neutral, less than 7 is acidic and more than 7 is alkaline. A *universal indicator* gives different colours corresponding to different pH values, typical results being red for pH = 1, purple for pH = 14.

12.2 Acids

An **acid** is a substance giving hydrated hydrogen ions, $H^+(aq)$, in aqueous solution (see also Section 12.5). Sometimes hydrated hydrogen ions are considered as H^+ ions joined to individual water molecules, in which case they are referred to as *oxonium ions*, H_3O^+. Acids show their normal properties only if water is present because these properties are due to $H^+(aq)$ ions, not to the molecules of the acids.

(a) Properties of Acids

1. Acids taste sour.
2. Acids have pH values less than 7 and turn blue litmus red.
3. With reactive metals (e.g. magnesium, zinc, iron) they give hydrogen.

$$\text{acid} + \text{metal} \rightarrow \text{salt} + \text{hydrogen}$$
$$2HCl(aq) + Mg(s) \rightarrow MgCl_2(aq) + H_2(g)$$
$$\text{or} \quad 2H^+(aq) + Mg(s) \rightarrow Mg^{2+}(aq) + H_2(g)$$

N.B. (i) Metals below iron in the reactivity series (see Section 13.1) do not give hydrogen with dilute acids.
(ii) Dilute nitric acid gives one or more of the oxides of nitrogen, not hydrogen, when reacted with metals.

4. With bases, neutralisation occurs to give a salt and water *only* (see Section 12.6).

$$\text{acid} + \text{base} \rightarrow \text{salt} + \text{water}$$
$$2HNO_3(aq) + CuO(s) \rightarrow Cu(NO_3)_2(aq) + H_2O(l)$$
$$\text{or} \quad 2H^+(aq) + O^{2-}(s) \rightarrow H_2O(l)$$

5. With carbonates (and hydrogencarbonates) effervescence occurs.

$$\text{acid} + \text{carbonate} \rightarrow \text{salt} + \text{water} + \text{carbon dioxide}$$
$$H_2SO_4(aq) + ZnCO_3(s) \rightarrow ZnSO_4(aq) + H_2O(l) + CO_2(g)$$
$$\text{or} \quad 2H^+(aq) + CO_3^{2-}(s) \rightarrow H_2O(l) + CO_2(g)$$

12.3 Bases and Alkalis

Bases are substances containing oxide ions, O^{2-}, or hydroxide ions, OH^-. They are usually oxides or hydroxides of metals.

Alkalis are soluble bases: they give hydrated hydroxide ions on dissolving in water. The common alkalis are potassium hydroxide, sodium hydroxide, calcium hydroxide and ammonia solution (often called ammonium hydroxide).

(a) Properties of Bases

1. Bases evolve ammonia (choking alkaline gas) from ammonium salts, especially on warming.

$$CaO(s) + (NH_4)_2SO_4(s) \rightarrow CaSO_4(s) + H_2O(l) + 2NH_3(g)$$
$$\text{or } O^{2-}(s) + 2NH_4^+(s) \rightarrow H_2O(l) + 2NH_3(g)$$

2. With acids, bases give a salt and water *only* (see properties of acids).

(b) Properties of Alkalis

Since alkalis are bases, they have the two properties given above. Their solutions have the following additional properties.

1. They feel soapy.
2. They have pH values greater than 7 and turn red litmus blue.
3. With aluminium and zinc they give hydrogen and an aluminate or a zincate.

12.4 Strengths of Acids and Bases

Acids with low pH values, such as hydrochloric, nitric and sulphuric acids are called *strong acids*. Similarly, alkalis with high pH values, such as solutions of potassium hydroxide and sodium hydroxide, are called *strong alkalis*. These substances are strong electrolytes (see Section 7.1) and are completely ionised in aqueous solution. *Weak* acids and alkalis have pH values nearer to 7: they are incompletely ionised in aqueous solution and are weak electrolytes. Examples include ethanoic acid and ammonia solution (see Section 7.1).

12.5 The Brønsted − Lowry Theory of Acids and Bases

According to this theory, an **acid** is a proton donor and a **base** is a proton acceptor.

These definitions include all of the familiar acids and bases but extend the range to other substances.

e.g.　　$HCl(aq) + H_2O(l) \rightleftharpoons H_3O^+(aq) + Cl^-(aq)$ . . . . (i)

In the forward reaction the water molecule accepts a proton from the hydrogen chloride molecule. The hydrogen chloride molecule is therefore an acid and the water molecule is a base. In the back reaction the oxonium ion is the acid and the chloride ion the base.

When ammonia reacts with water the equation is:

$NH_3(aq) + H_2O(l) \rightleftharpoons NH_4^+(aq) + OH^-(aq)$ . . . .(ii)
　base　　　　acid　　　　acid　　　　base

Notice that the water acts as a *base* in (i) and as an *acid* in (ii). Substances which show both acidic and basic properties are said to be **amphoteric**.

12.6 Salts

A **normal salt** is the product formed when all of the hydrogen ions of an acid are replaced by metal or ammonium ions, e.g. zinc chloride, $ZnCl_2$, is a salt of hydrochloric acid and sodium sulphate, Na_2SO_4, is a salt of sulphuric acid.

An **acid salt** is one in which only some of the hydrogen ions of an acid have been replaced by metal or ammonium ions. For example, in sodium hydrogensulphate, $NaHSO_4$, only half of the hydrogen ions in sulphuric acid have been replaced by sodium ions (see Example 12.5).

The number of hydrated hydrogen ions obtained from one molecule of an acid is called the **basicity** of the acid. Thus hydrochloric acid, HCl, has a basicity of one and is *monobasic*, sulphuric acid, H_2SO_4, has a basicity of two and is *dibasic*.

Neutralisation is the reaction between the hydrated hydrogen ions of an acid and the oxide or hydroxide ions of a base to form water, a salt being formed at the same time.

(a) Methods of Preparing Salts

There are three main methods of making salts and it is necessary to know whether a salt is soluble in water before the correct method can be chosen. Table 12.1 shows the solubilities of common salts and should be memorized.

Table 12.1

	Carbonates	All *insoluble* except potassium, sodium and ammonium carbonates.
Salts	Chlorides	All *soluble* except silver and lead chlorides. Lead chloride is soluble in hot water.
	Nitrates	All *soluble*.
	Sulphates	All *soluble* except barium and lead sulphates. Calcium sulphate is only slightly soluble.

Note: all common potassium, sodium and ammonium salts are soluble.

1. Insoluble salts are prepared by the *precipitation method*.
2. Potassium, sodium and ammonium salts are prepared by the *titration method*.
3. All other soluble salts are made by the *insoluble base method*.
4. In addition to the above methods, *direct synthesis* from the elements is sometimes used, especially for anhydrous chlorides.

Details of these methods are given in Example 12.3.

12.7 Worked Examples

Example 12.1

Hydrogen is produced from:
A Copper and aqueous sulphuric acid.

B Lead and aqueous nitric acid.

C Magnesium and aqueous sulphuric acid.

D Silver and aqueous hydrochloric acid.

E Zinc and aqueous nitric acid.

The answer is **C**.

See Sections 12.2 and 13.1.

Example 12.2

On warming with aqueous sodium hydroxide, compound X produced a gas which turned moist red litmus paper blue. The compound X could be

A aluminium nitrate

B ammonium nitrate

C calcium nitrate

D potassium nitrate

E zinc nitrate. (AEB, 1982)

The answer is **B**.

See Properties of Bases, Section 12.3.

Example 12.3

(a) Some of the ways in which *salts* may be prepared are (i) by synthesis, (ii) by reaction of acids with alkalis, (iii) by reaction of insoluble bases with acids and (iv) by precipitation.

Give **one** example of **each** of these four reactions, and for each name the reagents you would use, write an equation and state briefly how you would isolate a reasonably pure dry sample of the salt. (**4 × 5 marks**)

(b) Write an ionic equation for each of the reactions in (a) (ii) and (a) (iv), and explain the reaction in (a) (i) in terms of electron transfer. (**5 marks**)

(OLE)

(a) (i) *Anhydrous iron(III) chloride is made by synthesis, starting with iron filings and a supply of dry chlorine gas.*

$$2Fe(s) + 3Cl_2(g) \rightarrow 2FeCl_3(s)$$

The iron filings are heated in a combustion tube in a stream of dry chlorine. Brown fumes of iron(III) chloride sublime off and condense as dark brown crystals in a cool receiver.

(ii) *Sodium sulphate may be prepared by the reaction of an alkali with an acid (titration), starting with sodium hydroxide solution and dilute sulphuric acid.*

$$2NaOH(aq) + H_2SO_4(aq) \rightarrow Na_2SO_4(aq) + 2H_2O(l)$$

A known volume of sodium hydroxide solution is placed in a conical flask and a few drops of indicator such as litmus are added. Dilute sulphuric acid is added carefully from a burette, with stirring, until the indicator shows that the solution is neutral. The volume of acid added is noted. The solution is thrown away, the flask is rinsed with distilled water and the experiment is repeated using the same volumes of acid and alkali but this time without the indicator (which would contaminate the product).

The solution is evaporated until crystals form in a cooled drop and then the flask is left to cool. The resulting crystals of sodium sulphate are filtered off, washed with a little cold distilled water and dried between filter papers.

This method works for all potassium, sodium and ammonium salts *except* for the carbonates.

(iii) *Zinc nitrate may be prepared by the reaction of an insoluble base with an acid, starting with zinc oxide and dilute nitric acid.*

$ZnO(s) + 2HNO_3(aq) \rightarrow Zn(NO_3)_2(aq) + H_2O(l)$

Dilute nitric acid is warmed in a beaker and zinc oxide is added in small portions, with stirring, until excess remains undissolved in the bottom of the beaker. This shows that all of the acid has been neutralised. The excess solid is filtered off and dry crystals of zinc nitrate are obtained from the filtrate as described in the previous section.

This method will work using an insoluble metal carbonate, or even in some cases the metal itself, in place of the insoluble base. Substituting zinc carbonate or zinc powder for the zinc oxide would still give crystals of zinc nitrate.

(iv) *Silver chloride could be made by precipitation, starting with solutions of silver nitrate and sodium chloride.*

$AgNO_3(aq) + NaCl(aq) \rightarrow AgCl(s) + NaNO_3(aq)$

Equal volumes of solutions of the same concentration in mol dm^{-3} are mixed so that the mole ratio of the two solutes is the same as in the equation. The resulting white precipitate of silver chloride is filtered off, washed with a little distilled water and dried between filter papers.

This method can be used for *all insoluble salts or insoluble hydroxides*. The two starting materials are chosen so that each contains one of the ions which will join to form the precipitate. Both starting materials *must* be soluble. The other ions present *must not* combine to form a second precipitate as this would be impossible to separate from the required product, e.g. if silver sulphate and barium chloride solutions had been chosen in the above example, a mixture of silver chloride and barium sulphate would have been given.

$Ag_2SO_4(aq) + BaCl_2(aq) \rightarrow 2AgCl(s) + BaSO_4(s)$

If the *positive ion* which will go into the precipitate is always paired with *nitrate ions* and the *negative ion* from the precipitate with *sodium ions*, as in the original example, then soluble sodium nitrate will always be the second product and no complication will occur.

(b) $H^+(aq) + OH^-(aq) \rightarrow H_2O(l)$
 or $H_3O^+(aq) + OH^-(aq) \rightarrow 2H_2O(l)$
    $Ag^+(aq) + Cl^-(aq) \rightarrow AgCl(s)$

In (a) (i) three electrons are transferred from each iron atom to three chlorine atoms. The iron atoms are oxidised to Fe^{3+} ions and the chlorine atoms are reduced to Cl$^-$ ions.

Example 12.4

(a) A solution of a compound **R** in water (concentration 0.1 mol/dm^3) had a pH of 4 and reacted **slowly** with sodium carbonate to evolve a gas **T**.

    A solution of a compound **W** in water (concentration 0.1 mol/dm^3) had a pH of 1 and reacted **rapidly** with sodium carbonate to produce the same gas **T**.

  (i) Give the name of one substance in each case which would fit the description of **R**, **T** and **W**.

R could be ethanoic acid, T is carbon dioxide and W could be hydrogen chloride.

  (ii) Name the particle common to the solutions of **R** and **W** which is also responsible for the reactions of these solutions with sodium carbonate. Write an ionic equation for the reaction of this particle with a carbonate ion.

The particle is a hydrated hydrogen ion, $H^+(aq)$ (or oxonium ion, $H_3O^+(aq)$)

$$2H^+(aq) + CO_3^{2-}(s) \rightarrow H_2O(l) + CO_2(g)$$

or $\quad 2H_3O^+(aq) + CO_3^{2-}(s) \rightarrow 3H_2O(l) + CO_2(g)$

  (iii) Explain briefly why there is a difference in the rate of evolution of gas **T**. (Assume that the temperature was constant during the experiments.)   **(8 marks)**

Ethanoic acid is a weak acid and is not completely ionised in aqueous solution. Dilute hydrochloric acid is a strong acid and is completely ionised in aqueous solution. Thus the concentration of $H^+(aq)$ ions is greater in dilute hydrochloric acid and the reaction is faster.

(b) A solution containing 2.84 g of sodium sulphate was mixed with a solution containing 6.62 g of lead(II) nitrate, and a white precipitate **P** was formed; this precipitate was filtered, washed and dried. The filtrate was carefully concentrated, and colourless crystals **Q** were obtained on cooling.

  (i) Name the precipitate **P** and write a molecular equation for its formation.

P is lead(II) sulphate.

$$Pb(NO_3)_2(aq) + Na_2SO_4(aq) \rightarrow PbSO_4(s) + 2NaNO_3(aq)$$

  (ii) Calculate the maximum mass of **P** which could have been obtained.

mol $Pb(NO_3)_2$: mol Na_2SO_4 : mol $PbSO_4$ = 1 : 1 : 1 . . . . (A)

$$mol\ Pb(NO_3)_2\ used = \frac{6.62}{331} = 0.02$$

$$mol\ Na_2SO_4\ used = \frac{2.84}{142} = 0.02$$

$$\therefore mol\ PbSO_4\ formed = 0.02 \quad (from\ A)$$

$$\therefore g \quad PbSO_4\ formed = 0.02 \times 303$$

$$= 6.06$$

  (iii) Identify **Q** and calculate the maximum theoretical mass of **Q** which could have been obtained. Give **one** reason why the mass obtained in practice would have been less than the theoretical mass.   **(8 marks)**

                                   (AEB, 1982)

Q is sodium nitrate.

*Maximum amount of **Q** obtainable from 0.02 mol of $Pb(NO_3)_2$ and 0.02 mol of Na_2SO_4 is 0.04 mol (from equation).*

$\therefore$ Maximum mass = 0.04 × 85 g

                     = 3.4 g

In practice some sodium nitrate would be lost in the filtrate when the crystals are filtered off from the saturated solution in which they have grown.

Example 12.5

Sodium hydrogensulphate, $NaHSO_4$, is an acid salt. Its solution turns universal indicator paper red.

(a) What is meant by an 'acid salt'? **(1 mark)**

See definition in Section 12.6.

(b) What, roughly, is the pH of sodium hydrogensulphate solution? **(1 mark)**

1.

Sodium hydrogencarbonate solution turns universal indicator paper green.

(c) What is the approximate pH of sodium hydrogencarbonate solution? **(1 mark)**

8.

(d) Is sodium hydrogencarbonate an acid salt? Explain your answer. **(2 marks)**

Even though it is alkaline in solution, sodium hydrogencarbonate, $NaHCO_3$, is an acid salt because only half of the hydrogen ions in carbonic acid (carbon dioxide solution) have been replaced by sodium ions.

(e) What would you expect to see if solutions of sodium hydrogensulphate and sodium hydrogencarbonate were mixed? **(1 mark)**

The mixture would fizz.

Because acid + hydrogencarbonate gives carbon dioxide.

(f) Write an equation for the reaction. **(2 marks)**

$NaHSO_4(aq) + NaHCO_3(aq) \rightarrow Na_2SO_4(aq) + H_2O(l) + CO_2(g)$

Ionic equations for the reactions which occur when the two salts dissolve in water are

$HSO_4^-(aq) + H_2O(l) \rightleftharpoons H_3O^+(aq) + SO_4^{2-}(aq)$ (i)

$HCO_3^-(aq) + H_2O(l) \rightleftharpoons H_2CO_3(aq) + OH^-(aq)$ (ii)

(g) Define the terms 'acid' and 'base' according to the Brønsted–Lowry theory. **(2 marks)**

See definitions in Section 12.5.

(h) Which particles are acids and which are bases in equation (i)? **(2 marks)**

HSO_4^- *and* H_3O^+ *are acids;* H_2O *and* SO_4^{2-} *are bases.*

(i) Which particles are acids and which are bases in equation (ii)? **(2 marks)**

H_2O *and* H_2CO_3 *are acids;* HCO_3^- *and* OH^- *are bases.*

(j) Which word describes the behaviour of water in these two reactions? **(1 mark)**

Amphoteric.

12.8 Self-test Questions

Question 12.1

Hydrogen chloride dissolves in water and in dry methylbenzene (toluene).

(a) Use the following table to describe what will happen (if anything) when the reagents shown are added to each solution. **(6 marks)**

Reagent	Solution of hydrogen chloride in water	Solution of dry hydrogen chloride in dry methylbenzene
Dry neutral litmus paper		
Magnesium		
Anhydrous sodium carbonate		

(b) Explain why these two solutions behave differently. **(4 marks)**

(AEB, 1982)

Question 12.2

Which of the following compounds is a salt?

A Calcium oxide
B Lead(II) carbonate
C Sodium hydroxide
D Hydrogen chloride

(SEB)

Question 12.3

This question is about different methods of making nickel(II) nitrate, $Ni(NO_3)_2$.

Method I 25 cm^3 of 2M nitric acid, HNO_3, was placed in a beaker and green nickel(II) carbonate, $NiCO_3$, was added until it was in excess. After filtration green nickel(II) nitrate solution was obtained.

(a) Why was *excess* nickel(II) carbonate used? **(1 mark)**
(b) Why must the beaker be much larger than the volume of acid used? **(1 mark)**
(c) Write a balanced equation for the reaction, including state symbols. **(2 marks)**

Method II Nickel(II) oxide was dissolved in nitric acid according to the equation

$$NiO(s) + 2HNO_3(aq) \rightarrow Ni(NO_3)_2(aq) + H_2O(l)$$

(d) What volume of 2M nitric acid would be required to neutralise 7.5 g of nickel(II) oxide? (Relative atomic masses: O = 16, Ni = 59) **(3 marks)**

Method III Nickel(II) chloride solution was added to lead(II) nitrate solution

$$NiCl_2(aq) + Pb(NO_3)_2(aq) \rightarrow PbCl_2(s) + Ni(NO_3)_2(aq)$$

(e) Outline the main operation required to obtain nickel(II) nitrate crystals from 1.0 M nickel(II) chloride solution and 1.0 M lead(II) nitrate solution. **(2 marks)**

(L)

Question 12.4

Which of the following pairs of reagents will react to form a precipitate?

A Aqueous barium nitrate and aqueous sulphuric acid
B Aqueous sodium hydroxide and aqueous nitric acid
C Copper(II) carbonate and aqueous hydrochloric acid
D Aqueous calcium chloride and aqueous potassium nitrate

12.9 Answers to Self-test Questions

12.1 (a)

Reagent	Solution of hydrogen chloride in water	Solution of dry hydrogen chloride in dry methylbenzene
Dry neutral litmus paper	Turns red	No change
Magnesium	Effervescence of hydrogen	No change
Anhydrous sodium carbonate	Effervescence of carbon dioxide	No change

(b) Hydrogen chloride consists of covalent molecules. When it dissolves in water, it reacts to form $H^+(aq)$ (or $H_3O^+(aq)$) ions and $Cl^-(aq)$ ions, and the resulting solution is therefore acidic. It dissolves in methylbenzene without reacting and, as the resulting solution does not contain $H^+(aq)$ ions, it is not acidic.

The reactions in the middle column of the table are typical reactions of an acid.

$$2H^+(aq) + Mg(s) \rightarrow Mg^{2+}(aq) + H_2(g)$$
$$2H^+(aq) + CO_3^{2-}(aq) \rightarrow H_2O(l) + CO_2(g)$$

12.2 **B**.

D is an acid, the others are bases.

12.3 (a) To make sure that all of the dilute nitric acid had been used up.
  (b) The mixture will fizz and froth up, owing to the evolution of carbon dioxide.
  (c) $NiCO_3(s) + 2HNO_3(aq) \rightarrow Ni(NO_3)_2(aq) + H_2O(l) + CO_2(g)$
  (d) From the equation, mol NiO : mol HNO_3 = 1 : 2 . . . .(A)

$$g\ NiO = 7.5$$
$$mol\ NiO = \frac{7.5}{75}$$
$$mol\ HNO_3 = \frac{7.5}{75} \times 2 \quad (from\ A)$$
$$= 0.2$$

2 mol of HNO_3 are contained in 1 dm³ of 2M solution

(See Section 20.3).

∴ 0.2 mol of HNO_3 is contained in 0.1 dm³ of 2M solution
∴ Volume of acid required = 0.1 dm³ = 100 cm³

  (e) Mix equal volumes of the two solutions, filter off the precipitate of lead(II) chloride and obtain crystals of nickel(II) nitrate from the filtrate as described in part (a) (ii) of Example 12.3.

12.4 **A**.

The products are (A) barium sulphate (insoluble) and nitric acid, (B) sodium nitrate and water, (C) copper(II) chloride, carbon dioxde and water. In (D) there is no reaction.

13 The Metals

13.1 Introduction

When metals are arranged in order of decreasing chemical reactivity, the *reactivity series* is obtained. The chemical properties of a metal and its compounds are related to the position of the metal in the reactivity series as shown in Table 13.1.

Table 13.1

	Action of cold air	*Action of water*	*Action of dilute hydrochloric and sulphuric acids*	*Reduction of oxides*	*Nature of hydroxides*	*Effect of heat on carbonates*	*Effect of heat on nitrates*	
K	rapidly attacked	violent in cold	explode	not reduced by carbon or hydrogen	strongly basic: stable to heat	stable	give nitrite and oxygen	K
Na								Na
Ca	attacked	attacked in cold	give hydrogen with decreasing vigour as series is descended		weakly basic or amphoteric:	give oxide and carbon dioxide with increasing ease as series is descended	give oxide, nitrogen dioxide and oxygen with increasing ease as series is descended	Ca
Mg	little action: protective oxide coating formed	react with steam						Mg
Al				reduced by hot carbon	on heating give oxide and water with increasing ease as series is descended			Al
Zn								Zn
Fe	rusts in moist air			reduced by hot carbon or hydrogen				Fe
Pb	little action	no reaction	no reaction					Pb
Cu								Cu

(a) Displacement Reactions

Any metal will displace one lower in the reactivity series from aqueous solutions containing its ions. For example, if a piece of iron is placed in copper(II) sulphate solution, a reddish brown deposit of copper forms on the iron, and the blue colour of the solution fades. This is a redox reaction (see Section 4.3) in which the iron atoms are oxidised by each losing two electrons and going into solution as Fe^{2+} ions: Cu^{2+} ions are reduced by each gaining two electrons and forming a solid deposit of copper atoms (see Example 13.1 and Questions 13.1 and 13.2).

$$Fe(s) + Cu^{2+}(aq) \rightarrow Fe^{2+}(aq) + Cu(s)$$

A metal cannot displace another one above it in the reactivity series. Thus if a piece of copper were placed in iron(II) sulphate solution, no reaction would occur.

Hydrogen fits into the series between lead and copper. Magnesium ribbon will therefore displace hydrogen from aqueous solutions containing its ions (i.e. acids).

e.g. $Mg(s) + 2H^+(aq) \rightarrow Mg^{2+}(aq) + H_2(g)$

121

Since copper is below hydrogen in the series, it does not liberate the gas from dilute acids. In practice, lead does not do so either.

(b) Reduction of Metallic Oxides

A metallic oxide can be reduced to the metal by heating it with another metal which is higher in the reactivity series (see Example 13.2(d)).

e.g. $Zn(s) + PbO(s) \rightarrow ZnO(s) + Pb(s)$

Carbon fits into the series between aluminium and zinc. On heating, it will reduce the oxide of zinc and those metals below it, being itself oxidised to carbon monoxide or carbon dioxide.

Metals above zinc can reduce carbon dioxide on heating. For example, magnesium burns with a bright white light in carbon dioxide, being oxidised to a white ash of magnesium oxide. The carbon dioxide is reduced to black specks of carbon.

$2Mg(s) + CO_2(g) \rightarrow 2MgO(s) + C(s)$

Hydrogen behaves in this type of reaction as if it were *level* with iron. (Note that this is higher than its position for displacement reactions.) It will reduce the oxides of iron and metals below it, being itself oxidised to water. Iron and the metals above it reduce water to hydrogen and are themselves oxidised (see Table 13.1 and Example 13.4).

(c) Thermal Stability of Metallic Compounds

In general, the higher a metal is in the reactivity series, the less likely are its compounds to decompose on heating (see Table 13.1 and Question 13.2).

13.2 Potassium, Sodium and Lithium — the Alkali Metals

These are soft, silvery metals which are stored under oil because they rapidly tarnish in air. Lithium comes between sodium and calcium in the reactivity series.

In water, potassium melts and rushes around the surface as a silvery ball. Hydrogen is formed, and burns with a lilac flame owing to the presence of the potassium. The other product is potassium hydroxide solution.

$2K(s) + 2H_2O(l) \rightarrow 2KOH(aq) + H_2(g)$

Sodium reacts slightly less vigorously (the hydrogen does not inflame) and lithium less vigorously still.

The hydroxides of these metals are all strong alkalis.

Potassium nitrate and potassium sulphate are important fertilisers. Sodium hydroxide is an important industrial alkali, sodium carbonate is a water softener (see Section 13.3) and is used in glassmaking, while sodium hydrogen carbonate is an ingredient of baking powder and health salts. Sodium chloride is used in preparing food and in the manufacture of chlorine and sodium hydroxide in the mercury cell (see Section 18.3).

13.3 Calcium and Magnesium — the Alkaline Earth Metals

These are fairly reactive metals. Calcium reacts quite vigorously with cold water, producing hydrogen and a milky suspension of slightly soluble calcium hydroxide (see Example 13.4(b) and Question 13.3(f)).

$$Ca(s) + 2H_2O(l) \rightarrow Ca(OH)_2(aq) + H_2(g)$$

Magnesium metal reacts *very* slowly with cold water but burns when heated in steam to give magnesium oxide and hydrogen.

$$Mg(s) + H_2O(g) \rightarrow MgO(s) + H_2(g)$$

Calcium oxide (quicklime) is an important industrial base, made by heating calcium carbonate (chalk or limestone). It reacts violently with water to form calcium hydroxide, which is a cheap industrial alkali and can be used to neutralise soil acidity (see Example 13.5).

Magnesium metal is used in lightweight alloys, while magnesium hydroxide (Milk of Magnesia) reduces stomach acidity.

(a) Hardness of Water (see Example 13.6 and Question 13.4)

Hard water is water which does not readily form a lather with soap. Hardness is mostly caused by dissolved calcium or magnesium ions which react with soap to form an insoluble precipitate (scum) — see Section 19.7. Soapless detergents do not have insoluble calcium and magnesium salts, and therefore do not form scum.

Hardness which can be removed simply by boiling is called **temporary hardness.** It is due to dissolved magnesium hydrogencarbonate or calcium hydrogencarbonate, which decompose on heating to give a precipitate of the corresponding carbonate. How water becomes temporarily hard is described in Example 13.6(b).

Hardness which cannot be removed by boiling is called **permanent hardness** and is often due to dissolved sulphates of calcium and magnesium.

Methods of removing hardness ('softening' water) include:

1. Boiling — temporary hardness only.
2. Distillation — effective but expensive.
3. Adding sodium carbonate (washing soda) to precipitate insoluble magnesium carbonate or calcium carbonate.
4. Ion exchange. Some substances, called ion exchange resins, remove dissolved ions from hard water and replace them with ions of their own. If the cations (positive ions) are replaced with $H^+(aq)$ ions and the anions (negative ions) with $OH^-(aq)$ ions, these combine to form water molecules and the resulting product is very pure *deionised water*.

13.4 Aluminium

Aluminium is a silvery metal which, in spite of its high position in the reactivity series, is stable in air and water. This is because it is coated with a thin layer of oxide, which protects it from further attack. (See 'Anodising', Section 7.2(d).) It dissolves in dilute hydrochloric acid and in sodium hydroxide solution, liberating hydrogen in both cases.

Aluminium oxide and aluminium hydroxide are amphoteric (see Section 12.5).

The metal is used to make milk-bottle tops and food containers; it is also used in overhead electric cables and in light alloys. Its manufacture is described in Example 18.2.

13.5 Zinc

Zinc resembles aluminium to some extent, though its oxide coating is less protective. It gives hydrogen with dilute hydrochloric acid and with sodium hydroxide solution, and its oxide and hydroxide are amphoteric. Zinc oxide has the characteristic property of turning yellow on heating and back to white on cooling.

The metal is used to protect steel from rusting: the zinc coating (galvanising) corrodes instead of the steel. It is also used in roofing, to make the cases for dry batteries and to make alloys such as brass.

13.6 Iron

Iron is a member of the centre block of the periodic table (the transition metals). These metals have more than one valency, form coloured ions in solution and can act as catalysts. The rusting of iron is dealt with in Section 10.3 and Example 10.6.

When heated in air, iron forms magnetic iron oxide, Fe_3O_4. Heating in steam gives the same oxide, together with hydrogen. The other stable oxide is iron(III) oxide, Fe_2O_3. Iron(II) hydroxide (dirty green) and iron(III) hydroxide (reddish brown) are formed by precipitation when sodium hydroxide solution is added to a solution of an iron(II) salt or an iron(III) salt.

Iron(II) salts are generally pale green and less stable than the corresponding iron(III) salts, which are yellow or brown. Conversion of iron(II) compounds to iron(III) compounds involves loss of electrons and therefore is an example of oxidation. The reverse process is reduction.

$$Fe^{2+} \underset{\text{reduction}}{\overset{\text{oxidation}}{\rightleftharpoons}} Fe^{3+} + e^-$$

Cast iron is cheap but impure, hard and brittle; *wrought iron* is almost pure iron and is softer and more flexible. *Steels* are alloys of iron with carbon and other elements such as chromium and manganese.

The manufacture of iron and steel is described in Section 18.2 and Example 18.3.

13.7 Lead

Lead is a soft metal, affected only very slowly by air and water. Dilute hydrochloric and sulphuric acids have little action on it because it becomes coated with insoluble lead(II) chloride or lead(II) sulphate. Dilute nitric acid rapidly dissolves it, giving a solution of lead(II) nitrate and a mixture of the oxides of nitrogen.

There are three oxides of lead. Lead(II) oxide is an amphoteric orange solid, dissolving in dilute acids which have soluble lead(II) salts and in sodium hydroxide solution to give sodium plumbate(II).

e.g. $PbO(s) + 2HNO_3(aq) \rightarrow Pb(NO_3)_2(aq) + H_2O(l)$

Lead(IV) oxide is a brown powder which is a powerful oxidising agent. On heating it gives off oxygen and leaves a residue of orange lead(II) oxide.

$2PbO_2(s) \rightarrow 2PbO(s) + O_2(g)$

The third oxide of lead is red lead oxide, Pb_3O_4, which behaves as if it were a mixture of lead(II) oxide and lead(IV) oxide, i.e. $(2PbO + PbO_2)(s)$. The correct name for this compound is therefore dilead(II) lead(IV) oxide.

Lead(II) hydroxide is a white amphoteric solid, dissolving, like lead(II) oxide, in those dilute acids which have soluble lead(II) salts and in sodium hydroxide solution.

Apart from lead(II) nitrate and lead(II) ethanoate, the common lead(II) salts are insoluble in water. Lead(II) chloride is soluble in hot water but is deposited as crystals when the solution cools again.

Lead is used in roofing, in car batteries, for shielding against X-rays and radioactive material, and in alloys such as solder. Tetraethyl-lead(IV) (TEL) is added to petrol as an anti-knock compound, and consequently causes lead pollution of the air.

13.8 Copper

Copper, like iron, is a transition metal. It slowly tarnishes on exposure to air but is unaffected by water or steam. It is below hydrogen in the reactivity series and cannot displace the gas from dilute acids. Like lead, it dissolves in dilute nitric acid, giving a solution of blue copper(II) nitrate, together with nitrogen monoxide.

Copper(I) oxide is a red solid. Copper(II) oxide is a black basic solid.

Pale blue copper(II) hydroxide is insoluble in water but dissolves in ammonia solution to give a royal blue solution containing complex tetraamminecopper(II) ions.

Copper is used in roofing, as an electrical conductor and to make water pipes. It is alloyed with zinc to make brass and with tin to make bronze.

13.9 Differences between Metals and Non-Metals

See Example 13.4 and Table 4.1.

13.10 Worked Examples

Example 13.1

The following facts are known about four metals, P, Q, R and S.
  (i) R displaces P and S from solutions of their ions;
 (ii) Q reacts with water, R does not;
(iii) if plates of P and S are dipped into an electrolyte and joined by wires through a meter, electrons flow from P to S through the meter.

The order of activity, the most active being placed first, is

A R, Q, S, P
B Q, S, P, R
C P, S, Q, R
D Q, R, P, S

· (SEB)

The answer is **D**.

(i) tells us that R is above P and S in the reactivity series; (ii) tells us that Q is above R; (iii) tells us that P is forming positive ions, leaving behind electrons which flow to S. If P forms ions more readily than S, it must be more reactive than S.

Example 13.2

Refer to the following metals in answering this question:

aluminium   iron   potassium
copper   magnesium   zinc

(a) Name two metals which you would **not** use in the laboratory to prepare hydrogen from an appropriate aqueous acid. Give a reason in each case. **(4 marks)**

Metal: *copper.* Reason: *because it is below hydrogen in the reactivity series.*

Metal: *potassium.* Reason: *because the reaction is too violent.*

(b) (i) Name two metals which, though high in the 'reactivity series', do not tarnish or corrode noticeably in air. **(2 marks)**

Aluminium, zinc.

(ii) Explain why this is the case for one of the metals you have chosen. **(2 marks)**

Aluminium is coated with a thin layer of aluminium oxide, which protects it from further reaction with the air.

(c) Name one metal which forms common compounds in which the metal can exist in one of two oxidation (valency) states. Give the formulae of two of its compounds, one illustrating each state. **(3 marks)**

Name of metal: *iron.*

Formulae of compounds: $FeCl_2$ *and* $FeCl_3$.

(d) You are informed that a finely divided metal, when heated with the oxide of a metal lower in the 'reactivity series', will usually react highly exothermically.

(i) Write the equation for such a reaction. Both metals concerned should be selected from the list given. **(2 marks)**

$Mg(s) + CuO(s) \rightarrow MgO(s) + Cu(s)$

(ii) Suggest two practical safety measures which should be employed in demonstrating this reaction. **(2 marks)**

Use a small quantity of the mixture and use a safety screen or, better, a fume cupboard.

(e) 'Yellow metal' is an alloy containing 60% copper and 40% zinc by weight. Calculate the number of moles of zinc atoms in 260 g of the alloy (Zn = 65) **(3 marks)**

(NISEC)

$$g \text{ of zinc in } 260 \text{ g of alloy} = 260 \times \frac{40}{100} = 104$$

$$\text{Moles of atoms in } 104 \text{ g of zinc} = \frac{104}{65}$$

$$= 1.6$$

Example 13.3

The following experiment was carried out on a metal, M. Sufficient warm, concentrated nitric acid was added **just** to dissolve the metal. Brown fumes were given off and a colourless solution remained. A small piece of cleaned copper was then added to this solution. A precipitate formed and the copper dissolved.

(a) What were the brown fumes? **(1 mark)**

Nitrogen dioxide.

See Section 15.2.

(b) Which of the following could have been the metal M?

Zinc, silver, magnesium, iron. **(1 mark)**

Silver.

M must be below copper in the reactivity series since copper displaces it from solution.

(c) What would have happened to the metal M if it had been warmed with hydrochloric acid instead of nitric acid? Explain your answer. **(2 marks)**

(SEB)

Nothing would have happened because M is below hydrogen in the reactivity series and therefore cannot displace it from dilute acids.

Example 13.4

(a) Draw a table to illustrate the differences between the following properties of metals and non-metals:
  (i) physical appearance
  (ii) electrical properties
  (iii) chemical properties of their oxides
  (iv) type of bonding present in their chlorides. **(8 marks)**

Properties	Metals	Non-metals
Appearance	Shiny solids	Gases, liquids or dull solids
Electrical properties	Conduct electricity well when solid or molten	Do not conduct electricity
Chemical properties of oxides	Usually ionic, basic or amphoteric solids	Usually covalent acidic gases
Type of bonding in chlorides	Ionic	Covalent

(b) Compare the reactions, if any, of the metals calcium, copper and iron with (i) water or steam, (ii) dilute hydrochloric acid, and place the metals in order of decreasing reactivity.
  Write equations for any reactions that you describe. **(7 marks)**

(i) *Calcium reacts fairly vigorously with cold water.*

See Section 13.3 for details and equation.

Iron reacts on heating in steam to give magnetic iron oxide and hydrogen.

$$3Fe(s) + 4H_2O(g) \rightleftharpoons Fe_3O_4(s) + 4H_2(g)$$

Copper does not react with water or with steam, even on heating.

(ii) *Calcium reacts very vigorously with cold dilute hydrochloric acid, forming calcium chloride solution and hydrogen.*

$$Ca(s) + 2HCl(aq) \rightarrow CaCl_2(aq) + H_2(g)$$

Iron reacts fairly slowly with cold dilute hydrochloric acid, forming iron(II) chloride solution and hydrogen.

$$Fe(s) + 2HCl(aq) \rightarrow FeCl_2(aq) + H_2(g)$$

Copper does not react with dilute hydrochloric acid.
The order of reactivity of the three metals is thus calcium, iron, copper.

Example 13.5

(a) Name **three** naturally occurring forms of calcium carbonate. **(3 marks)**

Chalk, limestone, marble.

(b) How is calcium carbonate converted into
    (i) calcium oxide?   **(3 marks)**
    Give the equation for the reaction.

By heating strongly to drive off carbon dioxide.

$$CaCO_3(s) \rightarrow CaO(s) + CO_2(g)$$

    (ii) calcium hydroxide?   **(4 marks)**
    Give the equation for the reaction.

By adding water carefully to the calcium oxide formed in (i). Much heat is produced and the solid swells up to form a fine powder.

$$CaO(s) + H_2O(l) \rightarrow Ca(OH)_2(s)$$

(c) Discuss the part that calcium compounds play in
    (i) the extraction of iron   **(7 marks)**
    (ii) agriculture   **(4 marks)**
    (iii) the building industry.   **(3 marks)**
                                                    (AEB, 1981)

(i) *See Example 18.3.*

(ii) *Calcium carbonate and calcium hydroxide can be used to neutralise soil acidity. Calcium phosphate is used to make superphosphate fertiliser.*

(iii) *When limestone and clay are heated strongly together, cement is formed. Calcium sulphate is used in making plaster.*

Example 13.6

(a) Some fresh calcium oxide is exposed to the air and weighed at regular intervals. At first there is a fairly rapid increase in mass and this is followed by a much slower increase which eventually ceases. Explain these observations.   **(8 marks)**

The rapid increase in mass is caused by reaction between the calcium oxide and the moisture in the air to form calcium hydroxide.

$$CaO(s) + H_2O(g) \rightarrow Ca(OH)_2(s)$$

The much slower increase in mass which follows is due to reaction between the calcium hydroxide (alkaline) and carbon dioxide (acidic) in the air to form calcium carbonate. When this reaction is complete, no further change in mass occurs.

$$Ca(OH)_2(s) + CO_2(g) \rightarrow CaCO_3(s) + H_2O(g)$$

All alkaline hydroxides react in a similar way with carbon dioxide.

(b) Explain why
    (i) calcium carbonate is an essential ingredient of the material which goes into the blast furnace in the extraction of iron;
    (ii) calcium carbonate is insoluble in water, yet caves and potholes are common in limestone districts;
    (iii) calcium carbonate is insoluble in water, yet some hot-water pipes become blocked up by layers of this carbonate.   **(3 × 4 marks)**
                                                    (OLE)

(i) *See Example 18.3.*

(ii) *Rain water dissolves a little carbon dioxide as it falls through the air and much more as it soaks through the soil, where the gas is produced by living organisms. The resulting carbonic acid dissolves limestone (calcium carbonate) to form calcium hydrogencarbonate solution, giving rise to caves and potholes and making the water temporarily hard.*

$$CaCO_3(s) + H_2O(l) + CO_2(g) \rightleftharpoons Ca(HCO_3)_2(aq)$$

(iii) *The reaction shown by the above equation is reversible. When water containing temporary hardness is heated, the reaction goes backwards, causing layers of insoluble calcium carbonate to be built up in hot-water pipes and boilers.*

Example 13.7

A metal X forms a green chloride of formula XCl_2. It is likely that X
A is a transition metal
B forms a carbonate with the formula X_2CO_3
C is in the same group of the Periodic Table as magnesium
D reacts violently with cold water.
The answer is **A.**

Transition metals have coloured ions (see Section 13.6).

Example 13.8

Iron(III) oxide(s) $\xrightarrow{\text{A}}$ iron(s)

iron(s) $\xrightarrow{\text{B}}$ iron(II) chloride(aq) $\xrightarrow{\text{D}}$ iron(III) chloride(aq)

iron(II) chloride(aq) $\xrightarrow{\text{C}}$ iron(II) hydroxide(s)

(a) State the reagents and conditions needed for each of the above one-stage conversions, writing an equation for each reaction. **(9 marks)**
A *Heat strongly with carbon.*
$$Fe_2O_3(s) + 3C(s) \rightarrow 2Fe(s) + 3CO(g)$$
B *Dissolve in cold dilute hydrochloric acid.*
$$Fe(s) + 2HCl(aq) \rightarrow FeCl_2(aq) + H_2(g)$$

Heating iron in chlorine would give iron(III) chloride.

C *Add cold dilute sodium hydroxide solution.*
$$FeCl_2(aq) + 2NaOH(aq) \rightarrow Fe(OH)_2(s) + 2NaCl(aq)$$
D *Bubble chlorine gas into the solution in the cold.*
$$2FeCl_2(aq) + Cl_2(g) \rightarrow 2FeCl_3(aq)$$

(b) Why can reaction **D** be described as an electron transfer reaction? Illustrate your answer with an ionic equation. **(3 marks)**
Each Fe^{2+} ion loses an electron and is oxidised to an Fe^{3+} ion.
Each chlorine atom gains an electron and is reduced to a Cl^- ion.
Thus electrons are transferred from Fe^{2+} ions to chlorine atoms.
$$2Fe^{2+}(aq) + Cl_2(g) \rightarrow 2Fe^{3+}(aq) + 2Cl^-(aq)$$

Example 13.9

(a) Describe, giving essential practical details and an equation, how a sample of pure, dry lead(II) sulphate could be prepared, starting from lead(II) nitrate crystals. **(9 marks)**

The method is precipitation. The lead(II) nitrate crystals are dissolved in water and added to sodium sulphate solution. Details are given in Example 12.3 (a).

$Pb(NO_3)_2(aq) + Na_2SO_4(aq) \rightarrow PbSO_4(s) + 2NaNO_3(aq)$

(b) When dilead(II) lead(IV) oxide is heated, it decomposes according to the equation
$2Pb_3O_4 \rightarrow 6PbO + O_2$

(i) Calculate the mass of lead(II) oxide which could be obtained from 6.85 g of dilead(II) lead(IV) oxide, showing clearly how you arrive at the answer.

$2Pb_3O_4(s) \rightarrow 6PbO(s) + O_2(g)$

$\therefore\ 2 \times 685\ g\ Pb_3O_4\ give\ 6 \times 223\ g\ PbO$

$\therefore\ \ \ 6.85\ g\ Pb_3O_4\ give\ \dfrac{6 \times 223 \times 6.85\ g\ PbO}{2 \times 685}$

$= 6.69\ g\ PbO$

$M_r(Pb_3O_4) = 685;\ M_r(PbO) = 223$

(ii) Describe briefly one test to confirm that oxygen is evolved in this reaction.

If a glowing splint is held over the heated mixture, it will relight, confirming that oxygen is being evolved.

(iii) If lead(II) oxide is heated in a stream of dry hydrogen gas, what products will result? Write an equation for the reaction and explain why this can be considered to be an example of redox. **(11 marks)**

Lead and water will be formed.

$PbO(s) + H_2(g) \rightarrow Pb(s) + H_2O(g)$

The lead(II) oxide has been reduced (it has lost oxygen) and the hydrogen has been oxidised (it has gained oxygen). Therefore this reaction is an example of redox.

(c) Describe what is observed when aqueous hydrogen chloride is mixed with aqueous lead(II) nitrate, and write an equation for the reaction. **(4 marks)**

A white precipitate of lead(II) chloride is produced.

$Pb(NO_3)_2(aq) + 2HCl(aq) \rightarrow PbCl_2(s) + 2HNO_3(aq)$

13.11 Self-test Questions

Question 13.1

Magnesium reacts with aqueous hydrochloric acid, whereas metal X does not. From this, it would be expected that

A magnesium would displace X from a solution containing ions of X

B magnesium would react less readily with oxygen than X

C magnesium oxide would be more easy to reduce than the oxide of X

D magnesium carbonate would decompose more easily than the carbonate of X on heating.

Question 13.2

Metal X is higher in the activity series than hydrogen and metal Y is lower. It is probable that

A the nitrate of X is the more stable

B X is the less reactive to water

C Y displaces X from a solution of one of its salts

D Y corrodes in air more readily than X does.

Question 13.3

Use the following list of metals to answer the questions below: aluminium, calcium, copper, iron, lead, magnesium, potassium. Each metal may be used once, more than once or not at all.

Select **one** metal in each case which

(a) forms an amphoteric oxide. **(1 mark)**

(b) would **NOT** displace lead from aqueous lead(II) ions. **(1 mark)**

(c) reacts on heating with dry chlorine to form a black solid. **(1 mark)**

(d) will form compounds which impart a lilac (pink) colour to a bunsen flame. **(1 mark)**

(e) forms a chloride which is insoluble in cold water, but is soluble in hot water. **(1 mark)**

(f) will sink to the bottom of a beaker of cold water and react vigorously, liberating a colourless gas. **(1 mark)**

(AEB, 1983)

Question 13.4

The extent of temporary hardness of water may be estimated by titration of the hydrogencarbonate ions in solution with hydrochloric acid according to the equation

$$Ca(HCO_3)_2 + 2HCl \rightarrow CaCl_2 + 2H_2O + 2CO_2$$

(a) In an experiment a 100 cm^3 sample of tap water required 12.0 cm^3 of 0.05 M hydrochloric acid to give an end point using methyl orange indicator.

   (i) How many moles of HCl were added? **(3 marks)**

   (ii) How many moles of $Ca(HCO_3)_2$ were present in 100 cm^3 of tap water, assuming that no other carbonates or hydrogencarbonates were present? **(2 marks)**

   (iii) How many moles of $Ca(HCO_3)_2$ were there per dm^3 of tap water? **(3 marks)**

   (iv) What is the concentration of $Ca(HCO_3)_2$ in grams per dm^3? **(2 marks)**

   (v) Why is an indicator used to find the end point of this titration? **(3 marks)**

(b) (i) Describe how you could investigate whether a sample of tap water contained both temporary and permanent hardness. **(3 marks)**

   (ii) Explain, with ionic equations, how sodium carbonate removes both forms of hardness. **(3 marks)**

(c) Describe experiments you could carry out to prove that the scale in a kettle was mostly calcium carbonate. **(6 marks)**

(OLE)

13.12 Answers to Self-test Questions

13.1 A.

See Section 13.1.

13.2 A.

See Section 13.1.

13.3 (a) aluminium.

Or lead.

    (b) copper.

The metal must be *below* lead in the reactivity series.

(c) iron.

Iron(III) chloride is black, or dark brown.

(d) potassium.

See Section 20.2.

(e) lead.

See Section 13.7.

(f) calcium.

See Section 13.3.

13.4 (a) (i) 1 000 cm^3 of 0.05 M hydrochloric acid contains 0.05 mol of HCl (see Section 20.3).

$$\therefore 12.0 \text{ cm}^3 \text{ of acid contains } 0.05 \times \frac{12}{1000} \text{ mol of HCl}$$

$$= 0.0006 \text{ mol of HCl}$$

(ii) From the equation, mol HCl : mol Ca(HCO$_3$)$_2$ = 2 : 1 = 1 : $\frac{1}{2}$

$$\therefore \text{ mol Ca(HCO}_3)_2 = 0.0003$$

(iii) If 100 cm^3 of tap water contains 0.0003 mol of Ca(HCO$_3$)$_2$

$\therefore$ 1 dm^3 (1 000 cm^3) contains 0.003 mol of Ca(HCO$_3$)$_2$

(iv) Molar mass of Ca(HCO$_3$)$_2$ = 40 + 2(1 + 12 + 48) = 162 g mol^{-1}

$\therefore$ 1 dm^3 of tap water contains 0.003 $\times$ 162 = 0.486 g Ca(HCO$_3$)$_2$

i.e. Concentration of Ca(HCO$_3$)$_2$ = 0.486 g per dm^3

(v) An indicator is needed because there is no colour change, no precipitate and no obvious evolution of gas during the titration, and therefore no other way of telling where the end point occurs.

(b) (i) Place 50 cm^3 of distilled water in a conical flask, and 0.5 cm^3 of soap solution, stopper the flask and shake it vigorously. If a lather does not remain unbroken on the surface of the water for two minutes, add further 0.5 cm^3 portions of soap solution, shaking after each addition, until it does.

Repeat with 50 cm^3 of tap water in place of the distilled water. If the tap water requires more soap solution than the distilled water did, it is hard.

Boil another 50 cm^3 portion of tap water for a few minutes and then allow it to cool. Replace any water that has evaporated with distilled water so that the volume is again 50 cm^3, and test this with soap solution as before.

If the third sample needs the same volume of soap solution as the first, the hardness was all temporary and has disappeared on boiling. If it needs the same volume as the second sample, the hardness is all permanent and none has disappeared. If the result is somewhere between those of the first and second samples, some but not all of the hardness has disappeared, so both types must have been present.

(ii) Sodium carbonate removes both forms of hardness by precipitating calcium and magnesium ions as insoluble carbonates.

$$Mg^{2+}(aq) + CO_3^{2-}(aq) \rightarrow MgCO_3(s)$$
$$Ca^{2+}(aq) + CO_3^{2-}(aq) \rightarrow CaCO_3(s)$$

(c) Test for calcium ions and carbonate ions as described in Chapter 20.

14 Carbon

14.1 Carbon

Carbon is an element which exhibits allotropy.

Allotropy is the existence of an element in more than one form in the same state.

The condition is important in distinguishing allotropy from straightforward melting and boiling.

Carbon has two allotropes, diamond and graphite. Both consist of pure carbon which may be proved by burning equal masses of the two substances in excess oxygen when the same mass of carbon dioxide is obtained from each but no other product is formed. Diamond and graphite have widely differing physical properties, these differences being due to variations in the way the atoms are packed together (see Example 14.1).

(a) Reactions and Uses of Carbon

1. Diamond: used in gemstones, cutting tools, rock borers.
2. Graphite: used as a lubricant, a moderator in atomic reactors, an electrical conductor and in pencil 'leads' (mixed with clay).
3. Carbon is a reducing agent: when carbon is heated with the oxides of the less reactive metals, it reduces them to the metal and is itself oxidised to carbon dioxide

$$2CuO(s) + C(s) \rightarrow 2Cu(s) + CO_2(g) \qquad \text{(see Section 13.1)}$$

In industry, coke is used to extract a number of metals such as zinc and iron from their oxide ores.

14.2 Carbon Dioxide, CO_2

Carbon dioxide may be prepared by the action of almost any acid on any carbonate but the reaction most commonly employed is that between dilute hydrochloric acid and marble chips (calcium carbonate). The gas is passed through water to remove hydrogen chloride, and is collected by downward delivery as shown in Example 14.5.

$$CaCO_3(s) + 2HCl(aq) \rightarrow CaCl_2(aq) + H_2O(l) + CO_2(g)$$

(a) Test for Carbon Dioxide (see Example 14.4)

Carbon dioxide turns lime water milky.

$$Ca(OH)_2(aq) + CO_2(g) \rightarrow CaCO_3(s) + H_2O(l)$$

(b) Properties of Carbon Dioxide

1. Carbon dioxide is a colourless, odourless gas.
2. It is denser than air.
3. It dissolves slightly in water to give an acidic solution.

$$CO_2(g) + water \rightleftharpoons CO_2(aq)$$
$$CO_2(aq) + H_2O(l) \rightleftharpoons 2H^+(aq) + CO_3^{2-}(aq)$$

4. Carbon dioxide does not burn and will only support the combustion of substances which are hot enough to decompose it into its elements (see Example 14.2).

(c) Uses of Carbon Dioxide

1. In fizzy drinks (see Example 14.2).
2. In fire extinguishers (see Example 14.2).
3. As a refrigerant.
4. Health salts and baking powder contain sodium hydrogencarbonate and a solid acid such as tartaric acid which react together when water is added to produce carbon dioxide.

(d) Manufacture of Carbon Dioxide

Carbon dioxide is manufactured by heating limestone (calcium carbonate). It is stored under pressure in cylinders.

$$CaCO_3(s) \rightarrow CaO(s) + CO_2(g)$$

14.3 Carbonates

Potassium, sodium and ammonium carbonates are soluble in water, all others being insoluble. The insoluble carbonates can be obtained as precipitates by mixing together solutions containing the required metal ions and carbonate ions.

$$ZnSO_4(aq) + Na_2CO_3(aq) \rightarrow ZnCO_3(s) + Na_2SO_4(aq)$$

Soluble carbonates are prepared from carbon dioxide and the corresponding alkali.

$$CO_2(g) + 2KOH(aq) \rightarrow K_2CO_3(aq) + H_2O(l)$$

All carbonates react with acids to give a salt, water and carbon dioxide (see Example 14.5).

$$CuCO_3(s) + H_2SO_4(aq) \rightarrow CuSO_4(aq) + H_2O(l) + CO_2(g)$$

Carbonates of metals low down in the reactivity series split up on heating to give the metal oxide and carbon dioxide (see Question 14.4).

$$PbCO_3(s) \rightarrow PbO(s) + CO_2(g)$$

14.4 Hydrogencarbonates

The only common solid hydrogencarbonates are those of potassium and sodium; calcium and magnesium hydrogencarbonates are found only in solution. All are prepared by passing excess carbon dioxide through a solution or a suspension of the corresponding hydroxide or carbonate.

$$K_2CO_3(aq) + H_2O(l) + CO_2(g) \rightleftharpoons 2KHCO_3(aq)$$

Hydrogencarbonates give carbon dioxide with acids and when heated.

$$NaHCO_3(s) + HCl(aq) \rightarrow NaCl(aq) + H_2O(l) + CO_2(g)$$
$$2NaHCO_3(s) \rightarrow Na_2CO_3(s) + H_2O(l) + CO_2(g)$$

14.5 The Carbon Cycle

The percentage of carbon dioxide in the atmosphere remains fairly constant. The ways in which carbon atoms circulate in nature are shown in the carbon cycle (see Example 14.6).

14.6 Carbon Monoxide, CO

Carbon monoxide is a product of incomplete combustion of, for example, coal or petroleum.

(a) Properties of Carbon Monoxide

1. Carbon monoxide is a colourless, odourless gas.
2. It is slightly less dense than air.
3. It is almost insoluble in water.
4. Carbon monoxide burns in air with a blue flame to form carbon dioxide.

$$2CO(g) + O_2(g) \rightarrow 2CO_2(g)$$

5. It reduces the oxides of the less active metals on heating (see Example 14.5).

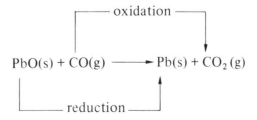

6. It is extremely poisonous (see Example 14.5).

14.7 Worked Examples

Example 14.1

Carbon is an element which exhibits allotropy. The structures of the two allotropes, diamond and graphite, are shown in Fig. 14.1.

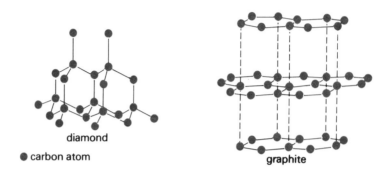

Fig. 14.1

(a) Explain what is meant by the term 'allotropy'. **(2 marks)**

Allotropy is the existence of an element in more than one form in the same state.

(b) By considering the structures of the allotropes, explain why

    (i) graphite conducts electricity but diamond does not. **(2 marks)**

In diamond, each carbon atom is using its four bonding electrons to form covalent bonds to four other atoms. In graphite, each carbon atom is bonded to only three others, even though four electrons are available for bonding. The extra electrons are free to move from one hexagon to the next within a layer and thus conduct an electric current.

    (ii) graphite is soft but diamond is hard. **(2 marks)**

Graphite is made up of layers of hexagons of carbon atoms. The bonding in each layer is strong but the layers are only held together by weak van der Waals' forces. The layers of carbon atoms can thus slide easily over one another. However, in diamond all the bonds are very strong.

    (iii) graphite has a much lower density than diamond. **(2 marks)**

Although both allotropes consist entirely of carbon atoms, graphite has a more open structure than that of diamond and hence has a lower density.

    (iv) both substances have high melting points. **(2 marks)**

This is because they both consist of giant structures of atoms (see Section 4.2).

Example 14.2

(a) (i) Give an equation for a reaction in which carbon dioxide is behaving as an oxidising agent. **(1 mark)**

$2Mg(s) + CO_2(g) \rightarrow 2MgO(s) + C(s)$

> Carbon dioxide oxidises the magnesium to white magnesium oxide, itself being reduced to black carbon.

(ii) Mention, and briefly explain, two uses of carbon dioxide in everyday life. **(4 marks)**
Carbon dioxide is used in fire extinguishers. Carbon dioxide does not burn and will only support the combustion of substances which are hot enough to decompose it into its elements, e.g. magnesium. The compressed gas may itself be directed on to the fire to blanket the flames (carbon dioxide is more dense than air) or it may be used to eject water, foam or powder from the extinguisher.
Carbon dioxide is used to make fizzy drinks. The carbon dioxide is dissolved under pressure; removing the top from the bottle releases this pressure, and bubbles of gas come out of solution.

(b) Carbon monoxide reacts with iron to give iron carbonyl, $Fe(CO)_5$. How many grams of iron carbonyl could be obtained from 1 dm^3 of carbon monoxide, measured at room temperature and pressure? **(5 marks)**

$Fe(s) + 5CO(g) \rightarrow Fe(CO)_5(l)$
$mol\ CO(g) : mol\ Fe(CO)_5(l) = 5 : 1 = 1 : 1/5\ ....A$
$\qquad\qquad dm^3\ CO(g) = 1$
$\qquad\qquad mol\ CO(g) = 1/24$ *(molar volume = 24 $dm^3\ mol^{-1}$ at room temperature and pressure)*

$mol\ Fe(CO)_5(l) \qquad = \dfrac{1}{24} \times \dfrac{1}{5}$ *(from A)*

$g\ Fe(CO)_5(l) \qquad = \dfrac{1}{24} \times \dfrac{1}{5} \times 196$ *($M_r\ (Fe(CO)_5) = 196$)*

$\qquad\qquad\qquad = 1.63$

Example 14.3

Carbon dioxide is not produced:
A during photosynthesis
B when copper(II) carbonate is heated
C when dilute hydrochloric acid is added to marble chips
D during the fermentation of glucose solution
E when oxygen is passed over heated diamonds.
The answer is **A.**

> Carbon dioxide is used up during photosynthesis, not produced.

Example 14.4

Marble chips are added to hydrochloric acid in a test tube.
(a) What do you observe? **(1 mark)**
Fizzing as a gas is given off.

(b) What is the chemical name for marble? **(1 mark)**
Calcium carbonate.

(c) Name two other forms of this chemical which occur in nature.   (2 marks)
Chalk and limestone.

(d) What do you observe if carbon dioxide is passed into lime water for a few seconds?   (1 mark)

The lime water will go milky.

(e) What is the chemical name for lime water?   (2 marks)
Calcium hydroxide solution.

(f) Write a word equation for the reaction.   (2 marks)
Calcium hydroxide + carbon dioxide → calcium carbonate + water

(g) Write a symbol equation for the reaction.   (2 marks)
$$Ca(OH)_2(aq) \quad + \quad CO_2(g) \quad \rightarrow \quad CaCO_3(s) \quad + H_2O(l)$$

(h) What do you observe if carbon dioxide is passed into lime water for several minutes?   (2 marks)

The lime water goes milky and then clear again.

(i) There would have been little reaction if sulphuric acid had been added to marble chips. Explain.   (2 marks)
Calcium sulphate is produced and as this is insoluble it forms a layer around the marble chips and stops the reaction.

Example 14.5

(a) (i) Describe how you could prepare and collect a pure sample of carbon dioxide, starting with marble chips. You may draw a diagram if you wish.
Dilute hydrochloric acid is added to marble chips. The gas given off is passed through water to remove acid spray and is then collected by downward delivery (Fig. 14.2).

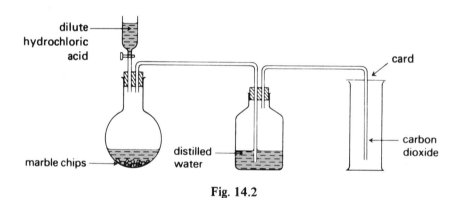

Fig. 14.2

(ii) Give the equation for the reaction.
$$CaCO_3(s) + 2HCl(aq) \rightarrow CaCl_2(aq) + H_2O(l) + CO_2(g)$$

(iii) State two ways in which the rate of reaction could be increased.   (6 marks)
The rate could be increased by using more concentrated acid or smaller pieces of marble (see Section 9.1).

(b) (i) In many places calcium hydrogencarbonate is present in tap water. Explain why this happens.
Rain water contains dissolved carbon dioxide (see Example 13.6). If this water passes through chalk or limestone regions, then it will dissolve the calcium carbonate to form calcium hydrogencarbonate.
$$CaCO_3(s) + H_2O(l) + CO_2(g) \rightleftharpoons Ca(HCO_3)_2(aq)$$

(ii) Explain the formation of stalactites and stalagmites in caves through which such water drips. **(6 marks)**

The slow evaporation of drops of dilute calcium hydrogencarbonate solution hanging from cave roofs causes the above process to reverse, and this leaves minute deposits of calcium carbonate behind. As a result of this, stalactites form. Where drops of solution fall to the floor and evaporate, stalagmites grow upwards.

(c) (i) Give details of an experiment to prove that carbon monoxide is a reducing agent.
Oxides of the less active metals, e.g. copper(II) oxide, are readily reduced by passing carbon monoxide over the heated metal oxide in the apparatus shown in Fig. 14.3.

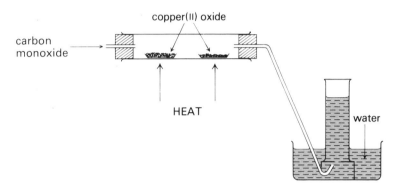

Fig. 14.3

The gas collected in the gas jar is shown to be carbon dioxide by testing it with lime water which goes milky. The black copper(II) oxide turns to reddish-brown copper and thus reduction of the copper(II) oxide must have occurred.
$$CuO(s) + CO(g) \rightarrow Cu(s) + CO_2(g)$$

(ii) Explain how breathing in carbon monoxide interferes with the ability of the haemoglobin in the blood to act as an oxygen carrier. **(8 marks)**

Oxygen combines with haemoglobin and is carried round the blood stream as bright red oxyhaemoglobin, which is unstable and readily gives up its oxygen where required. Carbon monoxide combines with haemoglobin to form cherry-red carboxyhaemoglobin. This is much more stable than oxyhaemoglobin and thus the blood can no longer act as an oxygen carrier.

Example 14.6

The ways in which carbon atoms circulate in nature are shown in the carbon cycle (Fig. 14.4).

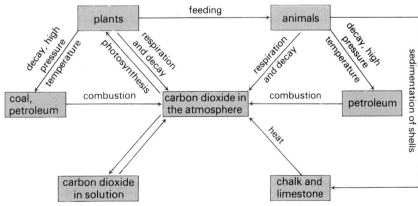

Fig. 14.4

(a) (i) Give two ways in which carbon dioxide is added to the air.

Carbon dioxide is added through combustion of substances such as coal and petroleum and through respiration.

(ii) Give two ways in which carbon dioxide is removed from the air.

Photosynthesis and the dissolving of carbon dioxide in water remove the gas from the atmosphere.

(iii) Explain why the percentage of carbon dioxide in the air stays approximately constant.   **(4 marks)**

The percentage of carbon dioxide in the air stays approximately constant owing to the delicate balance between the processes evolving the gas and those absorbing it.

(b) A white precipitate is formed when
(i) carbon dioxide is blown into lime water;
(ii) temporarily hard water is boiled;
(iii) washing soda is added to water containing dissolved magnesium chloride.
In each case name the precipitate and write an equation.   **(6 marks)**

                                                                                                (SUJB)

(i) *Calcium carbonate is formed.*
$$Ca(OH)_2(aq) + CO_2(g) \rightarrow CaCO_3(s) + H_2O(l)$$
(ii) *Again, calcium carbonate (or magnesium carbonate) is formed (see Section 13.3).*
$$Ca(HCO_3)_2(aq) \rightarrow CaCO_3(s) + H_2O(g) + CO_2(g)$$
(iii) *Magnesium carbonate is the precipitate.*
$$MgCl_2(aq) + Na_2CO_3(aq) \rightarrow MgCO_3(s) + 2NaCl(aq)$$

Example 14.7

Carbon is able to form a large number of compounds because it:
A is the most abundant element in the earth's crust
B can join onto other carbon atoms to form chains
C exists in two allotropic forms
D forms carbon dioxide which is present in the atmosphere
E occurs in living things.

The answer is **B**.

Example 14.8

Petrol is a mixture of hydrocarbons of the type C_nH_{2n+2}.
(a) Explain the presence of the following gases in the exhaust fumes of a car (equations are **not** expected as part of your answer):
(i) carbon dioxide;   **(2 marks)**

Complete combustion of a hydrocarbon produces carbon dioxide and water vapour (see Section 19.2).

(ii) carbon monoxide.   **(2 marks)**

If insufficient oxygen is present, then incomplete combustion occurs and carbon monoxide is formed.

(b) Explain the origins of the following substances which are also present in the exhaust fumes:
(i) lead(II) oxide and lead(II) bromide;   **(2 marks)**

Tetraethyl-lead(IV) is added to petrol to prevent premature ignition of the mixture of petrol and air in the cylinder of an engine. When burned, this compound produces lead(II) oxide.

To prevent a build-up of lead in the engine, 1,2-dibromoethane is added to the petrol. From this, lead(II) bromide is formed.

      (ii) oxides of nitrogen such as NO and NO_2.   **(2 marks)**
                                                       (OLE)

Oxides of nitrogen are produced when the oxygen and nitrogen in the air combine at the high temperatures produced in the engine.

14.8 Self-test Questions

Question 14.1

Complete the following equations:
(a) zinc carbonate(s) + hydrochloric acid(aq) → _____ + _____ + _____ .
(b) $ZnCO_3(s)$ → _____ + _____ .
(c) Copper oxide(s) + carbon monoxide(g) → _____ + _____ .
(d) $Ca(OH)_2(aq) + CO_2(g)$ → _____ + _____ .   **(10 marks)**

Question 14.2

A correct statement about charcoal, graphite, diamond and ethanol is that they all:
A are allotropes of carbon
B have high melting points
C are insoluble in water
D form carbon dioxide on combustion
E decompose on heating to form carbon.   (L)

Question 14.3

What type of reaction is occurring in each of the examples given below? Choose from the list **A–D** for your answer (in some cases two letters are needed).
A thermal decomposition
B precipitation
C neutralisation
D oxidation
(a) The formation of barium sulphate from barium chloride solution and sulphuric acid.
(b) The formation of carbon monoxide from carbon.
(c) Bubbling carbon dioxide into lime water for a few seconds.
(d) The action of heat on copper carbonate.   **(5 marks)**

Question 14.4

Complete the following sentences:
(a) Diamond and graphite resemble each other chemically because.........................
(b) When carbon dioxide is bubbled into lime water (aqueous calcium hydroxide), a white precipitate is obtained but this redissolves with excess carbon dioxide because...............
(c) When copper(II) carbonate is heated, it turns black because...................   **(6 marks)**

14.1 (a) Zinc chloride(aq), water(l) and carbon dioxide(g).
   (b) ZnO(s) and CO_2(g).
   (c) Copper(s) and carbon dioxide(g).
   (d) $CaCO_3$(s) and H_2O(l).

14.2 **D.**

The equations are

$C(s) + O_2(g) \rightarrow CO_2(g)$
$C_2H_5OH(l) + 3O_2(g) \rightarrow 2CO_2(g) + 3H_2O(l)$

14.3 (a) **B.**
   (b) **D.**
   (c) **B and C.**
   (d) **A.**

14.4 (a) They are both allotropes of carbon *or* they contain atoms of one kind only.
   (b) Calcium carbonate (the white precipitate) reacts further to give calcium hydrogencarbonate which is soluble.

$CaCO_3(s) + CO_2(g) + H_2O(l) \rightleftharpoons Ca(HCO_3)_2(aq)$

   (c) Black copper(II) oxide is formed.

$CuCO_3(s) \rightarrow CuO(s) + CO_2(g)$

---

15 Nitrogen and Phosphorus

15.1 Nitrogen, N_2

Nitrogen forms about 78% of the air and can be prepared from it by removing oxygen and carbon dioxide. Industrially, nitrogen is obtained by the fractional distillation of liquid air (see Section 10.2).

(a) Properties of Nitrogen

1. Nitrogen is a colourless, odourless gas.
2. It is a little less dense than air.
3. It is almost insoluble in water.
4. Nitrogen is generally unreactive. However, it forms nitrogen monoxide when sparked with oxygen, and reacts with hydrogen to give ammonia (see Section 18.5).

$$N_2(g) + O_2(g) \rightleftharpoons 2NO(g)$$
$$N_2(g) + 3H_2(g) \rightleftharpoons 2NH_3(g)$$

(b) Uses of Nitrogen

1. In the manufacture of ammonia.
2. To provide an inert atmosphere, e.g. in flushing out oil tanks.
3. As a refrigerant.

15.2 The Oxides of Nitrogen

Nitrogen monoxide, NO, is a colourless gas which is immediately oxidised to brown fumes of nitrogen dioxide, NO_2, on contact with oxygen.

$$2NO(g) + O_2(g) \rightleftharpoons 2NO_2(g)$$

Both oxides are emitted in car exhaust fumes, causing pollution of the air.

15.3 Ammonia, NH_3

Ammonia is prepared by warming any ammonium salt with any base (see Question 15.1).

$$2NH_4Cl(s) + Ca(OH)_2(s) \rightarrow CaCl_2(s) + 2H_2O(g) + 2NH_3(g)$$

The gas is dried by passing it over calcium oxide and is collected by upward delivery (see Example 15.2). Ammonia is manufactured by the Haber process (see Section 18.5).

(a) Tests for Ammonia

Ammonia has a characteristic choking smell and will turn universal indicator paper purple. It gives white fumes of ammonium chloride when mixed with hydrogen chloride (see Example 15.3).

$$NH_3(g) + HCl(g) \rightleftharpoons NH_4Cl(s)$$

(b) Properties of Ammonia

1. Ammonia is a colourless gas with a pungent choking smell.
2. It is about half as dense as air.
3. Ammonia is very soluble in water, giving an alkaline solution.

   $$NH_3(g) + water \rightleftharpoons NH_3(aq)$$
   $$NH_3(aq) + H_2O(l) \rightleftharpoons NH_4^+(aq) + OH^-(aq)$$

   Since it contains hydroxide ions, ammonia solution neutralises acids (giving ammonium salts) and will also precipitate insoluble metallic hydroxides from solutions containing the metal ions.
4. Air will oxidise ammonia if a platinum catalyst is present (see Section 18.6).

$$4NH_3(g) + 5O_2(g) \longrightarrow 4NO(g) + 6H_2O(g)$$

$$\Big\updownarrow \text{oxygen}$$

$$4NO_2(g)$$

(c) Uses of Ammonia

1. To make nitrogenous fertilisers, e.g. ammonium sulphate, ammonium nitrate.
2. To make nitric acid by the Ostwald process (see Section 18.6).
3. Ammonia solution is used in cleaning, as a grease remover.

15.4 Ammonium Salts

Ammonium salts are soluble in water and are prepared by the titration method (see Section 12.6). They decompose on heating and some, such as ammonium chloride, sublime (see Question 15.1). They react with sodium hydroxide solution on warming to give ammonia, this reaction being used to test for their presence.

$$NH_4Cl(aq) + NaOH(aq) \rightarrow NaCl(aq) + NH_3(g) + H_2O(l)$$

15.5 Nitric Acid, HNO_3

(a) Properties of Nitric Acid

Dilute nitric acid shows the usual properties of a strong acid (see Section 12.2 and Question 15.3) but it is reduced by metals to one or more of the oxides of nitrogen and does not generally give off hydrogen.

(b) Uses of Nitric Acid

1. To make fertilisers, e.g. ammonium nitrate.
2. To make explosives, e.g. TNT.

15.6 Nitrates

All nitrates are soluble in water and are prepared by the titration method (sodium, potassium and ammonium) or by the insoluble base method (see Section 12.6). Section 12.6).

$$KOH(aq) + HNO_3(aq) \rightarrow KNO_3(aq) + H_2O(l)$$
$$PbO(s) + 2HNO_3(aq) \rightarrow Pb(NO_3)_2(aq) + H_2O(l)$$

All nitrates decompose on heating. Nitrates of metals at the top of the reactivity series (potassium, sodium) melt and split up to give the nitrite and oxygen.

$$2NaNO_3(l) \rightarrow 2NaNO_2(l) + O_2(g)$$

All other metal nitrates decompose to give the metal oxide, nitrogen dioxide and oxygen.

$$2Cu(NO_3)_2(s) \rightarrow 2CuO(s) + 4NO_2(g) + O_2(g)$$

(a) Test for Nitrates

If a nitrate is warmed with sodium hydroxide solution and aluminium powder is added, ammonia will be given off. (This can be detected by its smell, and by the fact that it turns moist red litmus paper blue.)

15.7 The Nitrogen Cycle

Nitrogen is essential to plants and animals for the production of protein. This element is recirculated in nature as shown in Example 15.6.

15.8 Proteins

Proteins are produced by living cells to build up animal and plant tissue. They are made from amino acids, molecules of which contain an amino group ($-NH_2$) and a carboxylic acid group ($-COOH$). All of them evolve ammonia when heated with soda lime (see Example 15.5).

15.9 Phosphorus

Like nitrogen, phosphorus is found in group V of the periodic table. However, it is more reactive than nitrogen, e.g. it will burn with a brilliant yellow flame in a limited air supply to produce phosphorus(III) oxide, P_4O_6, or phosphorus(V) oxide, P_4O_{10}, if the oxygen is in excess. White phosphorus is spontaneously inflammable in chlorine giving phosphorus trichloride, or phosphorus pentachloride if the chlorine is in excess. Red phosphorus is much less reactive.

Phosphates are needed for healthy plant and animal life.

15.10 Worked Examples

Example 15.1

(i) Calculate the percentage by mass of nitrogen in ammonium sulphate.　　　(4 marks)

$(NH_4)_2SO_4$

Relative molecular mass $= (14 + 4) \times 2 + 32 + (4 \times 16)$
　　　　　　　　$= 132$

Percentage of nitrogen $= \dfrac{28}{132} \times 100$
　　　　　　　　$= 21.2$

(ii) Both ammonium sulphate and sodium nitrate are used as fertilisers. Which is the faster acting? Explain your answer.　　　(4 marks)

Sodium nitrate. Most plants can take in nitrogen only in the form of nitrates. However, ammonium sulphate can be oxidised to a nitrate by bacteria and so can be used as a fertiliser.

Example 15.2

(a) A student was asked to show her classmates how to prepare a solution of ammonia gas in water, and she set up the apparatus shown in Fig. 15.1. State briefly **three** mistakes which she has made in her experiment.　　　(3 marks)

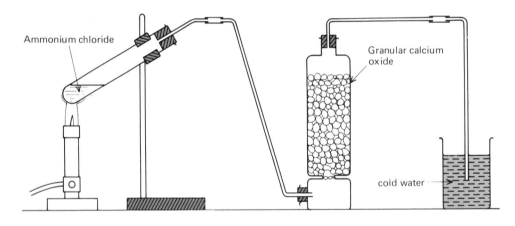

Fig. 15.1

Ammonia is produced by warming ammonium chloride with a base such as calcium hydroxide, not from ammonium chloride on its own.

$2NH_4Cl(s) + Ca(OH)_2(s) \rightarrow CaCl_2(s) + 2H_2O(g) + 2NH_3(g)$

The tube containing the two solids must be horizontal or sloping downwards. In the figure the tube slopes upwards: this means that the water produced would run back into the hot tube and crack it.

Ammonia is very soluble in water: 'sucking back' would result if it were dissolved by the method shown. The funnel arrangement must be used instead (see Example 18.6).

(b) Calculate the maximum mass of copper(II) oxide which could be reduced by 68 g of ammonia gas. **(4 marks)**

$3CuO(s) + 2NH_3(g) \rightarrow 3Cu(s) + 3H_2O(g) + N_2(g)$
mol $NH_3(g)$: *mol* $CuO(s) = 2 : 3 = 1 : 3/2$ . . . . .A
$\quad g\ NH_3\ = 68$
mol $NH_3\ = \dfrac{68}{17}\quad (M_r(NH_3) = 17)$
$\qquad\quad = 4$
mol $CuO\ = 4 \times 3/2\quad (from\ A)$
$\quad g\ CuO = 4 \times 3/2 \times 80\quad (M_r(CuO) = 80)$
$\qquad\quad\ = 480$

Example 15.3

The gases methylamine (CH_5N) and ammonia are closely related compounds and have very similar properties. Each can be represented by the formula $R–NH_2$.
  (i) State what R will be in the case of (1) ammonia, and (2) methylamine. **(4 marks)**
  (ii) Predict the effect of methylamine on (1) water coloured purple with litmus solution, and (2) hydrogen chloride gas. **(6 marks)**
    (OLE)

(i) *(1) R* = H
    (2) Methylamine has the formula CH_3NH_2.
    R = CH_3
(ii) *(1) The litmus solution will turn blue since methylamine is an alkaline gas.*

   (2) Methylamine will react with hydrogen chloride to give white fumes, similar to ammonium chloride.

   $CH_3NH_2(g) + HCl(g) \rightleftharpoons CH_3NH_3^+Cl^-(s)$
   cf. $NH_3(g) + HCl(g) \rightleftharpoons NH_4^+Cl^-(s)$

Example 15.4

(a) (i) What is the electron arrangement in a nitrogen atom? **(1 mark)**
2.5

   (ii) By means of a drawing, show the electron arrangement in a nitrogen molecule. **(1 mark)**

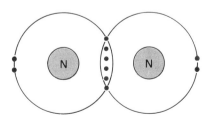

Fig. 15.2

147

(b) Ammonia burns in oxygen, forming nitrogen and water, according to the equation

$$4NH_3(g) + 3O_2(g) \rightarrow 2N_2(g) + 6H_2O(l)$$

If 80 cm³ of ammonia are burnt in 100 cm³ of oxygen, what will be the total volume of gas after combustion, and what will be its composition?

(Assume that all volumes are measured at room temperature and pressure.) **(4 marks)**

$4NH_3(g)$	+	$3O_2(g)$	$\rightarrow$	$2N_2(g)$	+	$6H_2O(l)$
4 molecules		*3 molecules*		*2 molecules*		*6 molecules*

By Avogadro's principle:

	4 volumes	*3 volumes*	*2 volumes*	—
i.e.	*80 cm³*	*60 cm³*	*40 cm³*	—

Volume of gas after combustion = 40 cm³ N₂ + (100 − 60) cm³ O₂
= 40 cm³ N₂ + 40 cm³ O₂
= 80 cm³

At room temperature, the water is present as a liquid and so its volume is negligible.

(c) (i) A jet of ammonia is spontaneously inflammable in chlorine, producing nitrogen and hydrogen chloride. Write an equation for this reaction. **(2 marks)**

$$2NH_3(g) + 3Cl_2(g) \rightarrow N_2(g) + 6HCl(g)$$

(ii) What will be observed if excess ammonia is used? **(2 marks)**

White fumes of ammonium chloride are seen since the hydrogen chloride produced will react with excess ammonia, i.e.

$$6NH_3(g) + 6HCl(g) \rightarrow 6NH_4Cl(s)$$

Adding the two equations together gives:

$$8NH_3(g) + 3Cl_2(g) \rightarrow N_2(g) + 6NH_4Cl(s)$$

Example 15.5

A substance was heated with soda lime. An alkaline gas was given off. The substance could have been

A A protein.
B A sugar.
C Cellulose.
D A vegetable oil. (SEB)

The answer is **A.**

Proteins are nitrogen containing compounds which react with soda lime to give ammonia.

Example 15.6

(i) What atmospheric conditions are needed for atmospheric nitrogen to be converted into nitrogen compounds in the soil (reaction 1 in Fig. 15.3)? **(1 mark)**

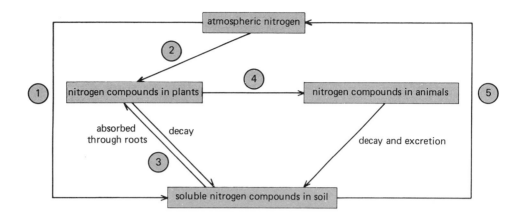

Fig. 15.3

A lightning flash.

> The energy in a lightning flash can cause atmospheric nitrogen and oxygen to combine to produce nitrogen monoxide. This combines with more oxygen to produce nitrogen dioxide which dissolves in water to enter the soil as nitric acid.

(ii) Some plants have bacteria in their roots and can absorb nitrogen directly from the atmosphere (reaction 2). Name one such plant. **(1 mark)**
Any leguminous plant, such as peas, beans or lupins, is a suitable answer.

(iii) What class of soluble nitrogen compound found in soil is absorbed through the roots of plants (reaction 3)? **(1 mark)**
Nitrates.

(iv) Name a type of nitrogen compound which is found in all plants and animals. **(1 mark)**
Proteins.

(v) How is reaction 4 brought about? **(1 mark)**
Plants are eaten by animals.

(vi) Suggest one way by which soluble nitrogen compounds in the soil are converted to atmospheric nitrogen (reaction 5). **(1 mark)**
Bacteria in the soil bring about this reaction.

(vii) Explain why it is necessary to add artificial fertilisers to the soil. **(2 marks)**
Man's removal of plants and animals from the land for use as food, together with modern methods of sewage disposal, result in insufficient nitrogen being returned to the soil by decay and excretion. It is essential to restore the balance using artificial fertilisers.

(viii) Name one nitrogen-containing compound used as a fertiliser. **(1 mark)**
Any one of ammonium nitrate, sodium nitrate or ammonium sulphate is a suitable answer.

(ix) Name two other elements which are required for healthy plant life. **(1 mark)**
Potassium and phosphorus.

15.11 Self-test Questions

Question 15.1

(a) The apparatus shown below was assembled and the ammonium chloride was gently heated. It was noticed that litmus paper A started to turn red and litmus paper B became blue; a white solid was seen to form near the cool, open end of the test-tube.

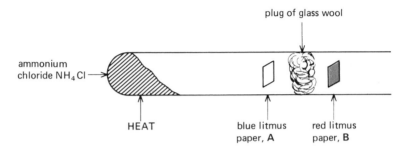

(i) Explain what happened to the ammonium chloride when it was heated and name the white solid formed near the open end of the test-tube. Write an appropriate equation.

(ii) Explain briefly why the atmosphere in the vicinity of A became acidic while in the vicinity of B it became alkaline. **(12 marks)**

(b) (i) Name a chemical which reacts with ammonium chloride to form ammonia gas, and write an equation for the reaction. Give the name of a suitable drying agent for ammonia and state how the gas might be collected after drying.

(ii) Calculate the volume of ammonia which could be obtained at room temperature and pressure from 10.7 g of ammonium chloride. **(8 marks)**

(c) Describe briefly **one** simple chemical test which would distinguish between aqueous ammonium chloride and aqueous ammonium sulphate. Write an equation for the reaction. **(4 marks)**

(AEB, 1981)

Questions 15.2–15.5

Directions. This group of questions deals with laboratory situations. Each situation is followed by a set of questions. Select the best answer for each question.

Questions 15.2–15.5 concern the preparation of nitric acid, using the simple distillation apparatus below. When sodium nitrate is mixed with concentrated sulphuric acid, an equilibrium is established:

$$NaNO_3(s) + H_2SO_4(l) \rightleftharpoons HNO_3(l) + NaHSO_4(s).$$

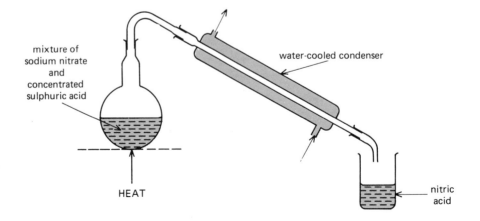

Nitric acid distils over when the mixture is heated. It is a fuming liquid with a boiling point of 83°C and a density of 1.5 g cm^{-3}. Sulphuric acid has a boiling point of 330°C and a density of 1.8 g cm^{-3}.

The 'nitric acid' obtained by this experiment is only 70% nitric acid by mass. Pure nitric acid is obtained by vacuum distillation. 63 g (1 mole) of pure nitric acid was added to water and then made up to 1000 cm^3 of solution and a rise in temperature of 4.8°C was observed. The dilute acid is a good conductor of electricity.
[4.2 kJ of heat energy raises the temperature of 1000 g of water by 1°C.]

Question 15.2

Nitric acid distils off from the equilibrium mixture in the flask when it is heated because
A nitric acid is the only substance in the mixture which is capable of being vaporized
B nitric acid has a lower density than sulphuric acid
C nitric acid has a lower boiling point than sulphuric acid
D sulphuric acid must remain behind to form $NaHSO_4$
E sulphuric acid is stronger than nitric acid   (L)

Question 15.3

Which of the following is a possible product of the reaction of dilute nitric acid with marble chips?
A Calcium
B Calcium oxide
C Calcium hydroxide
D Carbon dioxide
E Hydrogen   (L)

Question 15.4

The likely cause of the rise in temperature when the pure nitric acid is added to water is that
A the water is broken up into H^+ and OH^- ions
B the acid is broken up into H^+ ions and nitrate ions
C the acid combines with the water to form H_3NO_4 molecules
D the water takes protons from the acid to form $H_3O^+(aq)$ ions
E the acid is broken up by the water to give $H_2(g)$ (L)

Question 15.5

From the experimental observations, an approximate value of ΔH for the dilution of 1 mole of nitric acid is
A -4.2×4.8 kJ mol^{-1}

B $-\dfrac{4.2}{4.8}$ kJ mol^{-1}

C $-\dfrac{4.2}{63} \times 4.8$ kJ mol^{-1}

D $-\dfrac{4.2}{63} \times 4.8$ kJ mol^{-1}

E $-4.2 \times 4.8 \times 63$ kJ mol^{-1} (L)

15.12 Answers to Self-test Questions

15.1 (a) (i) On heating, ammonium chloride dissociates to give ammonia and hydrogen chloride, but these gases recombine on cooling. The white solid is thus ammonium chloride.

$$NH_4Cl(s) \rightleftharpoons NH_3(g) + HCl(g)$$

(ii) Ammonia, being less dense than hydrogen chloride, diffuses more quickly through the glass wool plug so that it will reach litmus paper B before hydrogen chloride does.

(b) (i) Copper(II) oxide.

In fact, any base will do.

$$2NH_4Cl(s) + CuO(s) \rightarrow CuCl_2(s) + 2NH_3(g) + H_2O(g)$$

The gas may be dried using potassium hydroxide pellets or calcium oxide and collected by upward delivery of air.

(ii) From the above equation,

mol NH_4Cl : mol NH_3 = 2 : 2 = 1 : 1 . . . . A

g NH_4Cl = 10.7

$$mol\ NH_4Cl = \frac{10.7}{53.5} \qquad (M_r(NH_4Cl) = 53.5)$$

$$mol\ NH_3 = \frac{10.7}{53.5} \times 1 \text{ (from A)}$$

$$dm^3\ NH_3 = \frac{10.7}{53.5} \times 24 \text{ (molar volume at room temperature and pressure = 24 dm}^3\text{ mol}^{-1})$$

$$= 4.8$$

(c) If barium chloride solution is added, aqueous ammonium sulphate gives a white precipitate of barium sulphate.

$$(NH_4)_2SO_4(aq) + BaCl_2(aq) \rightarrow BaSO_4(s) + 2NH_4Cl(aq)$$

15.2 **C.**

15.3 **D.**

Marble chips (calcium carbonate) react with nitric acid to give carbon dioxide.

$$CaCO_3(s) + 2HNO_3(aq) \rightarrow Ca(NO_3)_2(aq) + H_2O(l) + CO_2(g)$$

15.4 **D.**

Do not make the mistake of thinking that the answer is **B**. The nitric acid will break up into H^+ and nitrate ions but this is an endothermic process (bonds broken). The exothermic reaction that occurs is the hydration of the protons (bonds made), i.e.

$$H_2O(l) + HNO_3(aq) \rightarrow H_3O^+(aq) + NO_3^-(aq)$$

15.5 **A** (see Chapter 8).

16 Oxygen and Sulphur

16.1 Oxygen, O_2

Oxygen is the most abundant element in the earth's crust; it also forms 21% of the atmosphere.

Oxygen is prepared in the laboratory by adding 20 volumes hydrogen peroxide solution to manganese(IV) oxide as shown in Example 16.1.

$$2H_2O_2(aq) \rightarrow 2H_2O(l) + O_2(g)$$

It is a product of the electrolysis of dilute sulphuric acid (see Section 7.2).

Oxygen is manufactured by the fractional distillation of liquid air (see Section 10.2).

(a) Test for Oxygen

Oxygen relights a glowing splint.

(b) Properties of Oxygen

1. Oxygen is a colourless, odourless gas.
2. It is a little more dense than air.
3. It is slightly soluble in water.
4. Oxygen reacts with most elements to give their oxides (see Example 16.1), e.g. sodium burns with a yellow flame to give sodium oxide.

    $$4Na(s) + O_2(g) \rightarrow 2Na_2O(s).$$

    Sulphur burns with a bright blue flame to give mainly sulphur dioxide.

    $$S(s) + O_2(g) \rightarrow SO_2(g)$$

(c) Uses of Oxygen

1. In steel making to remove impurities from cast iron (see Section 18.2).
2. In oxy-acetylene blowpipes which are used for cutting and welding steel.
3. As an aid to breathing, e.g. in hospitals and climbing.
4. In rocket fuels.

16.2 Oxides

Oxides are compounds of oxygen with one other element, e.g. sodium oxide, Na_2O. There are several different types of oxide and these are discussed in Example 16.2.

16.3 Metal Hydroxides

A **metal hydroxide** is made up of metal ions and hydroxide ions (OH^-). Most metal hydroxides are insoluble in water and may be prepared by precipitation (see Section 12.6), e.g.

$$FeCl_3(aq) + 3NaOH(aq) \rightarrow Fe(OH)_3(s) + 3NaCl(aq)$$

Soluble hydroxides can be formed by the addition of the metal oxide to water (see Example 16.1).

All metal hydroxides dissolve in acids to give a salt and water only but the amphoteric hydroxides (e.g. aluminium, zinc and lead hydroxides) dissolve in alkalis as well (see Question 16.1).

$$Cu(OH)_2(s) + H_2SO_4(aq) \rightarrow CuSO_4(aq) + 2H_2O(l)$$

The hydroxides of potassium and sodium are stable to heat but those of metals below sodium in the reactivity series decompose into the corresponding oxide and water on heating, e.g.

$$Cu(OH)_2(s) \rightarrow CuO(s) + H_2O(l)$$

16.4 Sulphur

Sulphur is extracted from underground deposits by the Frasch process (see Example 16.3). It is also present as an impurity in crude oil and natural gas.

(a) Properties of Sulphur

1. Sulphur is a yellow, brittle, non-metallic solid with a relatively low melting point.
2. It is insoluble in water.
3. Sulphur is soluble in organic solvents, e.g. methylbenzene.
4. When heated, sulphur combines with most metals to form sulphides and with non-metals such as oxygen, chlorine and hydrogen, e.g.

$$Zn(s) + S(s) \rightarrow ZnS(s)$$
$$H_2(g) + S(s) \rightarrow H_2S(g)$$

(b) Uses of Sulphur

1. To make sulphuric acid.
2. To vulcanise (harden) rubber.

16.5 Sulphur Dioxide, SO_2

Sulphur dioxide is formed by burning sulphur and fuels containing sulphur compounds.

$$S(s) + O_2(g) \rightarrow SO_2(g)$$

(a) Properties of Sulphur Dioxide

1. Sulphur dioxide is a colourless, poisonous gas.
2. It has a characteristic choking smell.
3. It is about twice as dense as air.
4. Sulphur dioxide is very soluble in water. It reacts with water to give a solution of sulphurous acid, H_2SO_3.
5. It can be oxidised to sulphur trioxide.

$$2SO_2(g) + O_2(g) \rightleftharpoons 2SO_3(g)$$

(b) Uses of Sulphur Dioxide

1. As an intermediate in the production of sulphuric acid.
2. As a bleaching agent.
3. To preserve food.

16.6 Sulphuric Acid, H_2SO_4

Sulphuric acid is prepared by dissolving sulphur trioxide in concentrated sulphuric acid and then pouring the liquid into water (see Section 18.4).

(a) Properties of Concentrated Sulphuric Acid (see Example 16.5)

1. Sulphuric acid is a colourless, oily liquid. It is highly corrosive.
2. It is **hygroscopic** (it absorbs water vapour from the atmosphere). This makes it useful for drying gases (see Question 16.4). When poured into water, an exothermic reaction occurs.
3. It is a **dehydrating agent** (it can remove chemically combined water or the elements of water from other compounds) (see Example 16.5).
4. Concentrated sulphuric acid is a powerful oxidising agent (see Question 16.3).
5. Dilute sulphuric acid behaves as a typical strong acid (see Section 12.2, Example 16.5 and Question 16.3).

16.7 Sulphates

These are prepared as described in Section 12.6.

(a) Test for Sulphates

If dilute hydrochloric acid followed by a few drops of barium chloride solution is added to a sulphate solution, a white precipitate of barium sulphate is obtained, e.g.

$$Na_2SO_4(aq) + BaCl_2(aq) \rightarrow BaSO_4(s) + 2NaCl(aq).$$

16.8 Worked Examples

Example 16.1

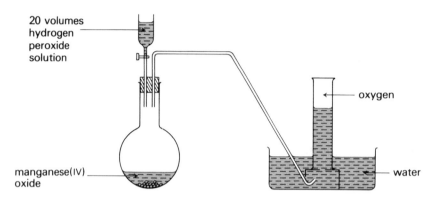

Fig. 16.1

Oxygen can be prepared by adding hydrogen peroxide solution to manganese(IV) oxide (Fig. 16.1). The manganese(IV) oxide is recovered unchanged in mass at the end of the reaction.

(a) What is the function of the manganese(IV) oxide in this reaction? **(1 mark)**

The manganese(IV) oxide acts as a catalyst.

Sodium and sulphur both burn in oxygen and the product of each reaction is soluble in water.

(b) Name the product formed when each element burns in oxygen. **(2 marks)**

Sodium oxide. Sulphur dioxide.

The equations for these reactions are

$$4Na(s) + O_2(g) \rightarrow 2Na_2O(s)$$
$$S(s) + O_2(g) \rightarrow SO_2(g)$$

(c) Give the names and approximate pH values of the two aqueous solutions. **(3 marks)**

Sodium hydroxide pH 13. Sulphurous acid pH 4.

The equations for these reactions are

$$Na_2O(s) + H_2O(l) \rightarrow 2NaOH(aq)$$
$$SO_2(g) + water \rightleftharpoons SO_2(aq)$$
$$SO_2(aq) + H_2O(l) \rightleftharpoons H_2SO_3(aq)$$

When zinc combines with oxygen the compound formed is insoluble in water but dissolves in both dilute hydrochloric acid and in sodium hydroxide solution.

(d) What is the word used to describe the behaviour of the zinc compound? **(1 mark)**

Amphoteric.

(e) Describe in outline the commercial preparation of oxygen from the air. **(3 marks)**

Oxygen is manufactured by the fractional distillation of liquid air (see Section 10.2).

(f) Give two large-scale uses of oxygen. **(2 marks)**

Choose any two of the uses in Section 16.1 for your answer.

Example 16.2

Classify the following oxides as acidic, basic, neutral or amphoteric:
  (i) ZnO;
  (ii) NO_2;
  (iii) MgO;
  (iv) Al_2O_3;
  (v) CO.

(5 marks)
(SUJB)

  (i) Amphoteric;
  (ii) acidic;
  (iii) basic;
  (iv) amphoteric;
  (v) neutral.

There are several types of oxide:
(a) Acidic oxides are usually the oxides of non-metals. They react with bases to form salts and with alkalis to form salts and water only. Many of them combine with water to form acids.
(b) Basic oxides are the oxides of metals. They react with acids to form salts and water only. If they dissolve in water they form alkalis.
(c) Neutral oxides are oxides of a few non-metals. They react with neither acids nor bases.
(d) Amphoteric oxides are the oxides of certain metals in the middle groups of the periodic table, e.g. aluminium, zinc and lead. They have the properties of both acidic and basic oxides, i.e. they react with both alkalis and acids to form salts and water only.

Example 16.3

(a) Sulphur is extracted from underground deposits by the Frasch process. Three concentric pipes, A, B and C, are sunk down to the deposits as shown in Fig. 16.2.

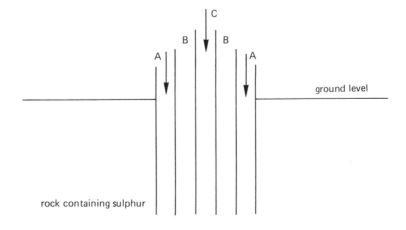

Fig. 16.2

Water at 170°C is pumped down the outer pipe.

(i) Explain how it is possible to obtain water with a temperature of 170°C.　**(1 mark)**

Water boils at 100°C at 1 atm pressure. If the pressure is increased, the boiling point of the water is raised.

(ii) What is the water used for?　**(1 mark)**

The water is used to melt the sulphur.

(iii) What is passed down tube C?　**(1 mark)**

Hot compressed air.

(iv) What happens in tube B?　**(1 mark)**

A froth of water, air and molten sulphur is forced up the middle pipe and is led off to settling tanks where the sulphur separates out.

(b) Give TWO large-scale uses of sulphur.　**(2 marks)**

To make sulphuric acid and to harden rubber.

(c) Given a finely powdered mixture of sulphur, salt and chalk, describe how you would recover a pure dry sample of each.　**(7 marks)**

Sulphur is soluble in methylbenzene. Methylbenzene is added to the mixture which is stirred well and then filtered. The filtrate (sulphur solution) is collected in an evaporating basin and the methylbenzene allowed to evaporate at room temperature.

Salt is soluble in water. The residue from the first part is added to warm water, stirred and filtered. The filtrate (salt solution) is evaporated to dryness (see Section 1.3).

The chalk is the residue on the filter paper. This can be washed with water and dried in a warm oven.

NB. You are asked to obtain a dry sample of each substance — it is not sufficient just to separate the solids.

Example 16.4

(a) Write down balanced equations to show how
(i) sulphur dioxide may form sulphur trioxide;　**(2 marks)**

$2SO_2(g) + O_2(g) \rightleftharpoons 2SO_3(g)$

(ii) sulphur trioxide forms sulphuric acid.　**(2 marks)**

$SO_3(s) + H_2O(l) \rightarrow H_2SO_4(aq)$

(b) Rain falling near some power stations is often slightly acidic. This 'acid' rain may be seen to affect some city statues and buildings but not others. Explain this observation.

(2 marks)
(OLE)

Coal-burning power stations generate sulphur dioxide which dissolves in water to form sulphurous acid and by oxidation sulphuric acid ('acid' rain). The 'acid' rain will attack those building materials which react with acids, e.g. marble.

Acid rain also has an effect on vegetation and animals.

Smokeless fuels generate less smoke and tar but produce more sulphur dioxide than the same mass of coal.

158

Example 16.5

Fig. 16.3

(a) What property of sulphuric acid does the hazard sign in Fig. 16.3 indicate? **(1 mark)**
It is corrosive.

(b) What are the hazards involved in the storage and transportation of sulphuric acid?
(3 marks)
Water must be excluded since the reaction between water and sulphuric acid is highly exothermic and hence dangerous. Sulphuric acid is a powerful oxidising agent and reacts vigorously with many substances, often producing sulphur dioxide which is a poisonous gas.

(c) Describe how sulphuric acid reacts with each of the following. State whether concentrated or dilute acid is used, and write an equation for the reaction.
    (i) copper(II) sulphate crystals.
Concentrated sulphuric acid converts blue crystals of copper(II) sulphate-5-water to the white anhydrous salt. It is behaving as a dehydrating agent.
$$CuSO_4 . 5H_2O(s) \rightarrow CuSO_4(s) + 5H_2O(l)$$

    (ii) sugar.
Concentrated sulphuric acid will remove the elements of water (i.e. hydrogen and oxygen atoms in the ratio of 2 : 1) from sugar. The mixture gets hot, swells up and leaves a black mass of carbon.
$$C_{12}H_{22}O_{11}(s) \rightarrow 12C(s) + 11H_2O(l)$$
The sulphuric acid is acting as a dehydrating agent here as well.

NB. Do not confuse dehydration with drying. A drying agent only removes water that is not chemically combined.

    (iii) magnesium. **(2 marks each)**
Dilute sulphuric acid is behaving here as a typical strong acid. Magnesium effervesces with dilute sulphuric acid to produce hydrogen.
$$Mg(s) + H_2SO_4(aq) \rightarrow MgSO_4(aq) + H_2(g)$$

16.9 Self-test Questions

Question 16.1

(a) Describe what happens if a few drops of sodium hydroxide solution are added to
    (i) copper(II)sulphate solution.
    (ii) zinc sulphate solution. **(2 marks)**
(b) What happens if excess sodium hydroxide solution is added? **(2 marks)**
(c) Explain your answer to (b). **(2 marks)**

Question 16.2

From the list
 A ammonia
 B chlorine
 C hydrogen
 D nitrogen
 E sulphur dioxide
choose the gas which is
(a) used to preserve food,
(b) prepared by the distillation of liquid air,
(c) soluble in water to give an alkaline solution,
(d) obtained as a by-product when fossil fuels are burned,
(e) used in the manufacture of margarine.   (1 mark each)

Question 16.3

Dilute sulphuric acid reacts with
 A copper to give copper(II) sulphate and hydrogen,
 B iron to give iron(II) sulphate and hydrogen,
 C magnesium carbonate to give magnesium oxide, carbon dioxide and sulphur dioxide.
 D zinc to give zinc sulphate and water,
 E zinc oxide to give zinc sulphate and hydrogen.

Question 16.4

Arrangements for drying a gas with concentrated sulphuric acid could include:

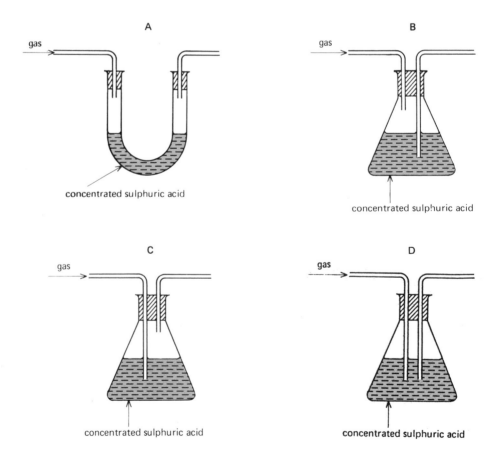

16.10 Answers to Self-test Questions

16.1 (a) (i) A blue precipitate of copper(II) hydroxide is formed.
(ii) A white precipitate of zinc hydroxide is formed.
(b) The copper(II) hydroxide remains unchanged but the zinc hydroxide dissolves.
(c) Zinc hydroxide is amphoteric and so reacts with excess alkali. Copper(II) hydroxide is basic.

16.2 (a) **E.**
(b) **D.**
(c) **A.**
(d) **E.**
(e) **C.**

16.3 **B.**

The relevant equations are
$Fe(s) + H_2SO_4(aq) \rightarrow FeSO_4(aq) + H_2(g)$
$MgCO_3(s) + H_2SO_4(aq) \rightarrow MgSO_4(aq) + CO_2(g) + H_2O(l)$
$Zn(s) + H_2SO_4(aq) \rightarrow ZnSO_4(aq) + H_2(g)$
$ZnO + H_2SO_4(aq) \rightarrow ZnSO_4(aq) + H_2O(l)$
Copper is below hydrogen in the reactivity series and so does not react with dilute sulphuric acid.
Concentrated sulphuric acid oxidises copper and is itself reduced to sulphur dioxide:
$Cu(s) + 2H_2SO_4(l) \rightarrow CuSO_4(s) + 2H_2O(l) + SO_2(g)$.

16.4 **C.**

17 The Halogens

17.1 Hydrogen Chloride, HCl

Hydrogen chloride is prepared in the laboratory by the action of concentrated sulphuric acid on rock salt (impure sodium chloride). It is manufactured by burning hydrogen in chlorine, both materials being obtained by the electrolysis of brine in the mercury cathode cell (see Section 18.3).

(a) Tests for Hydrogen Chloride

Hydrogen chloride forms steamy fumes in moist air. It will turn moist universal indicator paper red and gives a thick white precipitate of silver chloride with silver nitrate solution.

$$HCl(g) + AgNO_3(aq) \rightarrow AgCl(s) + HNO_3(aq)$$

(b) Properties of Hydrogen Chloride

1. Hydrogen chloride is a colourless gas with a pungent, choking smell.
2. It gives steamy fumes in moist air.
3. It is slightly more dense than air.
4. Hydrogen chloride is very soluble in water, forming hydrochloric acid. This is a strong acid (see Section 12.4 and also Example 17.1).
5. Hydrogen chloride reacts with ammonia to form thick white fumes of ammonium chloride.

$$NH_3(g) + HCl(g) \rightleftharpoons NH_4Cl(s)$$

6. Concentrated hydrochloric acid is easily oxidised to chlorine, e.g. by manganese(IV) oxide.

17.2 Chlorides

All of the common metallic chlorides are soluble in water, except those of silver and lead. They are prepared as described in Section 12.6 (see also Examples 17.3 and 17.4).

$$Pb(NO_3)_2(aq) + 2NaCl(aq) \rightarrow PbCl_2(s) + 2NaNO_3(aq)$$
$$CuO(s) + 2HCl(aq) \rightarrow CuCl_2(aq) + H_2O(l)$$

(a) Test for Chlorides

Chlorides react with silver nitrate solution in the presence of dilute nitric acid to give a white precipitate of silver chloride (see Example 17.1).

$$NaCl(aq) + AgNO_3(aq) \rightarrow AgCl(s) + NaNO_3(aq)$$

17.3 Chlorine, Cl_2

Chlorine is prepared in the laboratory by the oxidation of concentrated hydrochloric acid by a strong oxidising agent such as manganese(IV) oxide. The gas is passed through water to remove unchanged hydrogen chloride, dried by means of concentrated sulphuric acid, and collected by downward delivery.

$$MnO_2(s) + 4HCl(aq) \rightarrow MnCl_2(aq) + Cl_2(g) + 2H_2O(l)$$

Chlorine is manufactured by the electrolysis of sodium chloride solution using titanium anodes and a flowing mercury cathode (see Section 18.3 and Example 18.4).

(a) Test for Chlorine

Chlorine bleaches moist universal indicator paper.

(b) Properties of Chlorine

1. Chlorine is a greenish-yellow gas with a characteristic pungent, choking smell.
2. It is poisonous.
3. It is about twice as dense as air.
4. Chlorine is slightly soluble in water. It reacts with water to give a mixture of hydrochloric and chloric(I) acids (see Example 17.3). Hence, chlorine water is acidic.
5. Chlorine combines with most metals, non-metals and with hydrogen to give the corresponding chloride, particularly on heating (see Example 17.3).

$$2Na(s) + Cl_2(g) \rightarrow 2NaCl(s)$$
$$H_2(g) + Cl_2(g) \rightarrow 2HCl(g)$$

6. Chlorine is an extremely powerful oxidising agent (see Example 17.3).

(c) Uses of Chlorine

1. In the manufacture of plastics (e.g. PVC), anaesthetics, insecticides (e.g. DDT), solvents and aerosol propellants, all of which are chlorinated organic compounds.
2. As a bleach in the pulp and textile industries.
3. In the treatment of sewage and in the purification of water.

17.4 A Comparison of the Halogens

Atoms of the halogens all have 7 electrons in their outermost shells and therefore behave similarly (see Example 17.7).

Chlorine, bromine and iodine can all be obtained by the oxidation of the corresponding hydrogen halide.

The melting points and boiling points increase down the group so that fluorine is a yellow gas, chlorine a greenish-yellow gas, bromine a reddish-brown liquid and iodine a black solid. All the halogens combine with a large number of metals and non-metals and with hydrogen.

$$2Fe\,(s) + 3Br_2\,(g) \rightarrow 2FeBr_3\,(s)$$

In general, the reactions of fluorine are the most vigorous and those of iodine are the least vigorous. For example, bromine water like chlorine water is acidic but only about 1% of the dissolved bromine actually reacts with the water. Iodine is virtually insoluble in water but does show slight acidic properties by dissolving in sodium hydroxide solution.

The size of the atoms increases from fluorine to iodine (see Question 17.1). Thus fluorine which has the smallest atoms is the best electron attractor (i.e. oxidising agent) of these four elements and iodine the feeblest. Once the extra electron has been gained it will be more firmly held in a small ion such as fluoride than in a large ion. Thus it is possible for a change such as

$$Cl_2\,(g) + 2I^-\,(aq) \rightarrow I_2\,(aq) + 2Cl^-\,(aq)$$

to occur (see Example 17.6).

17.5 Worked Examples

Example 17.1

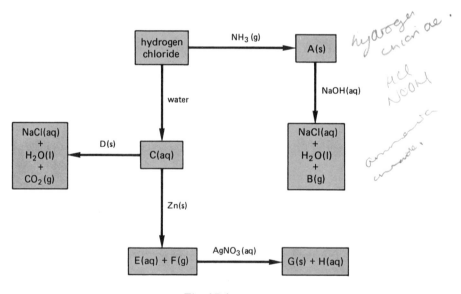

Fig. 17.1

(a) Name the compounds A to H.   **(8 marks)**
Hydrogen chloride gas reacts with ammonia to give white fumes of ammonium chloride (A).

$$NH_3(g) + HCl(g) \rightleftharpoons NH_4Cl(s)$$

Ammonium chloride reacts with sodium hydroxide solution to give ammonia (B).

$$NH_4Cl(s) + NaOH(aq) \rightarrow NaCl(aq) + H_2O(l) + NH_3(g)$$
This is a test for ammonium salts.

C is hydrochloric acid.
D is sodium carbonate.

$$Na_2CO_3(s) + 2HCl(aq) \rightarrow 2NaCl(aq) + H_2O(l) + CO_2(g)$$

Zinc reacts with hydrochloric acid to give zinc chloride (E) and hydrogen (F).

$$Zn(s) + 2HCl(aq) \rightarrow ZnCl_2(aq) + H_2(g)$$

Silver nitrate solution reacts with zinc chloride solution to give a white precipitate of silver chloride (G) and a solution of zinc nitrate (H).

$$ZnCl_2(aq) + 2AgNO_3(aq) \rightarrow 2AgCl(s) + Zn(NO_3)_2(aq)$$

  (b) Describe a test by which you could identify
      (i) gas B.
Ammonia turns universal indicator paper purple (see Section 15.3).

      (ii) gas F.   (2 marks)
Hydrogen 'pops' when a flame is applied (see Section 11.4).

Example 17.2

  (a) What are the essential processes involved in the purification of rock salt to give common
      salt?   (3 marks)
  (i) *Solution.*
 (ii) *Filtration.*
(iii) *Evaporation.*

  (b) What is the chemical name for common salt?   (1 mark)
Sodium chloride.

  (c) Give one advantage and one disadvantage of including common salt in our diet.
                                                                     (2 marks)
Advantage: *It is an essential mineral for health OR it improves the flavour of our food.*
Disadvantage: *It can lead to high blood pressure.*

  (d) Draw a diagram to show how chlorine may be obtained by the electrolysis of brine in
      the laboratory (Fig. 17.2).   (3 marks)

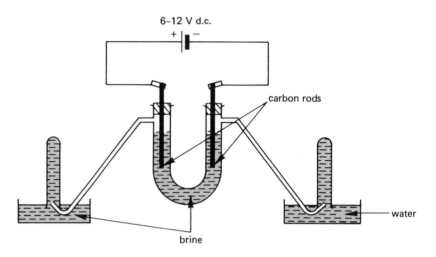

Fig. 17.2

(e) Write an electrode equation for the change taking place at the anode.   **(1 mark)**

$$2Cl^-(aq) - 2e \rightarrow 2Cl(g)$$
$$\downarrow$$
$$Cl_2(g)$$

Example 17.3

(a) The hazard signs shown in Fig. 17.3 are associated with chlorine gas. What do they mean?   **(2 marks)**

(i)   (ii)

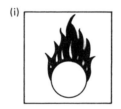   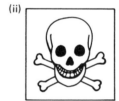

Fig. 17.3

(i) *Oxidising.*
(ii) *Toxic.*

(b) What reactions take place when chlorine is
(i) bubbled into cold water;   **(2 marks)**
When chlorine is passed into cold water, some of it simply dissolves but some of it reacts to give hydrochloric acid and chloric(I) acid.
$Cl_2(g) + water \rightleftharpoons Cl_2(aq)$
$Cl_2(aq) + H_2O(l) \rightleftharpoons HCl(aq) + HOCl(aq)$

(ii) passed over heated iron wool;   **(2 marks)**
Iron wool inflames in chlorine to give thick brown fumes of iron(III) chloride.
$2Fe(s) + 3Cl_2(g) \rightarrow 2FeCl_3(s)$

(iii) mixed with hydrogen and exposed to diffused sunlight?   **(2 marks)**
The two gases combine to produce hydrogen chloride.
$H_2(g) + Cl_2(g) \rightarrow 2HCl(g)$

(c) How would you distinguish between a solution in water of the salt prepared in (b, ii) and the product formed when iron filings are added to dilute sulphuric acid? Write an appropriate equation for any one reaction you use. **(2 marks)**

Sodium hydroxide solution produces a precipitate of red/brown iron(III) hydroxide with iron(III) chloride solution but a precipitate of green iron(II) hydroxide when added to iron(II) sulphate solution.

$$FeCl_3(aq) + 3NaOH(aq) \rightarrow 3NaCl(aq) + Fe(OH)_3(s)$$
$$red/brown$$
$$FeSO_4(aq) + 2NaOH(aq) \rightarrow Na_2SO_4(aq) + Fe(OH)_2(s)$$
$$green$$

(d) The addition of chlorine water to aqueous iron(II) sulphate results in the appearance of a yellow colour and the disappearance of the chlorine smell. Explain this reaction. **(2 marks)**

Chlorine is an oxidising agent, oxidising green Fe^{2+} to yellow Fe^{3+}. The chlorine molecules are reduced to chloride ions.

The ionic equation is
$$2Fe^{2+}(aq) + Cl_2(aq) \rightarrow 2Fe^{3+}(aq) + 2Cl^-(aq)$$

Example 17.4

Which one of the following methods would be most suitable for preparing a sample of anhydrous iron(II) chloride?
A Dissolving iron filings in aqueous hydrochloric acid followed by evaporating to dryness
B Dissolving iron(II) oxide in aqueous hydrochloric acid followed by evaporating to dryness
C Passing dry chlorine gas over heated iron
D Passing dry hydrogen chloride gas over heated iron
E Precipitation by mixing aqueous iron(II) sulphate and aqueous barium chloride.
(AEB, 1981)

The answer is **D.**

$$Fe(s) + 2HCl(g) \rightarrow FeCl_2(s) + H_2(g)$$
C would give iron (III) chloride.

Example 17.5

(a) How would you show that an aqueous solution contained
    (i) chlorine?
A piece of litmus paper placed in the solution will be bleached if chlorine is present.

    (ii) chloride ions?
If silver nitrate solution acidified with nitric acid is added to the solution, a white precipitate of silver chloride is formed if chloride ions are present. **(4 marks)**

(b) What is the difference between a chlorine atom and a chloride ion? **(1 mark)**
A chloride ion is charged and is formed by the addition of an electron to a chlorine atom.

Example 17.6

(a) Chlorine is obtained commercially by the electrolysis of fused sodium chloride.
  (i) Write down the equation for the production of chlorine at the appropriate elect-
    rode, and name the type of chemical reaction taking place. **(5 marks)**

$$2Cl^- (l) - 2e \rightarrow 2Cl(g)$$
$$\downarrow$$
$$Cl_2(g)$$

Chlorine is liberated at the anode (see Section 7.2). Since electrons are lost,
oxidation is taking place.

  (ii) Give **two** important uses of chlorine. **(2 marks)**
Choose any two uses in Section 17.3 as your answer.

(b) Explain the chemistry of the bleaching action of moist chlorine. **(4 marks)**
When chlorine is passed into water, a yellow–green solution consisting of dissolved
chlorine, hydrochloric acid and chloric(I) acid is formed.
$$Cl_2(g) + water \rightleftharpoons Cl_2(aq)$$
$$Cl_2(aq) + H_2O(l) \rightleftharpoons HCl(aq) + HOCl(aq)$$
The chloric(I) acid is unstable and readily gives up its oxygen to dyes, thereby
converting them to colourless products.
$$dye + HOCl(aq) \rightarrow (dye + oxygen) + HCl(aq)$$
coloured   colourless

(c) Describe a reaction of chlorine with a metallic element. Your description should include
    a statement of the conditions under which the reaction occurs. **(3 marks)**
Chlorine reacts with strongly heated iron to give brown fumes of iron(III) chloride
(see Example 17.3).

(d) Bromine and iodine are members of the same group of the periodic table as chlorine.
    Describe tests by which these three elements could be arranged in an order of activity.
    (6 marks)
    (OLE)

If chlorine water is added to an aqueous solution of sodium bromide containing a
little 1,1,1-trichloroethane, then the chlorine, being more reactive than bromine,
will displace it from the solution. The bromine produces an orange coloration
with 1,1,1-trichloroethane. A similar experiment using bromine water and an
aqueous solution of sodium iodide produces a violet coloration in the 1,1,1-tri-
chloroethane due to the displacement of iodine.
  Thus the order of activity is:

                    chlorine
                    bromine
                    iodine.

Example 17.7

(a) Fluorine, chlorine, bromine and iodine are all in Group VII of the Periodic Table. What
    does this tell you about:
  (i) the electronic structures of atoms of bromine and iodine;
Both atoms will have 7 electrons in their outermost shells.

  (ii) the molecular formula of bromine and iodine;
The molecular formulae of bromine and iodine will be similar to that of chlorine,
i.e. Br_2 and I_2.

  (iii) the formula of sodium bromide;
NaBr *(similar to* NaCl*).*

  (iv) the solubility of silver iodide? **(4 marks)**

It will be insoluble in water (like silver chloride).

    (b) Explain why fluoride ions are readily formed from fluorine atoms.   **(2 marks)**
                                                                                        (OLE)

A fluorine atom is relatively small and so its nucleus readily attracts an extra electron.

17.6 Self-test Questions

Question 17.1

Which one of the following statements is true?
A a chlorine atom is larger than a bromine atom
B a chlorine atom is larger than a chloride ion
C a bromide ion is larger than a bromine atom
D a fluorine atom is larger than an iodine atom
E a fluoride ion is larger than an iodide ion.

Question 17.2

What would you observe when chlorine is bubbled into potassium iodide solution?
                                                                          (2 marks)
                                                                          (SUJB)

Question 17.3

    (a) The element iodine is in the same Group of the Periodic Table as chlorine and bromine. Use this information to complete the following table: **(3 marks)**

	Formula	*Solid, liquid or gas at room temperature?*	*Soluble or insoluble in water?*
Potassium iodide	(i)	(ii)	(iii)
Hydrogen iodide	(iv)	(v)	(vi)

    (b) For the reaction

        $Cl_2 + 2I^- \rightarrow 2Cl^- + I_2$

    state which species has been oxidised and explain your answer.   **(2 marks)**
                                                                                 (OLE)

17.7 Answers to Self-test Questions

17.1 C.

> As a group in the Periodic Table is descended, further shells of electrons are added and so the atoms become bigger. Halide ions are larger than the corresponding atoms. When a further electron is added to an atom, the attraction of the nucleus for the electrons remains the same but the mutual repulsion between the electrons increases; hence the radius becomes larger.

17.2 A brown coloration would develop in the solution.

Chlorine is more reactive than iodine so it will displace it from solutions containing its ions.

$$Cl_2(aq) + 2I^-(aq) \rightarrow 2Cl^-(aq) + I_2(aq)$$

Iodine is soluble in potassium iodide solution, producing a brown solution.

17.3 (a) (i) KI
           (ii) solid
          (iii) soluble
           (iv) HI
            (v) gas
           (vi) soluble.

(b) The I^- ions have been oxidised since they have each lost an electron to form an iodine atom.

18 Some Industrial Processes

18.1 Introduction

Industrial processes are designed to manufacture products as economically as possible, taking into account both the speed and the cost of operation.

Chemical works should ideally be sited near the source of raw materials and the users of the products, or near to a good transport system. Control of pollution is an important factor to take into account, even though it can add a considerable amount to the cost involved.

In recent years the realisation that the earth's resources are not limitless has led to an increase in the recycling of materials. For example, scrap steel and aluminium can be reprocessed and made into new items.

18.2 Extraction of Metals

The method of extracting a metal from its compounds depends upon the position of the metal in the reactivity series. Electrolysis is the only economic way of reducing ions of metals near the top of the series: for example, aluminium is obtained by electrolysing a solution of aluminium oxide in molten cryolite (see Example 18.2). Metals such as this could not be manufactured cheaply until the means of generating large amounts of electricity had been invented.

For metals lower down in the series, chemical reduction is used. The commonest reducing agent is coke (impure carbon, made by heating coal in the absence of air). Iron is manufactured by heating its ore, often impure iron(III) oxide, with limestone and coke in a blast furnace (see Example 18.3).

The blast furnace produces *cast iron,* which is hard and brittle owing to the presence of impurities such as carbon, sulphur and phosphorus. Steel is stronger and more flexible than iron. It is made by removing the impurities from cast iron and then adding carbon and other elements such as chromium in the correct proportions to give the required product.

Oxygen is blown on to the surface of molten cast iron in a converter of the type shown in Fig. 18.1. It oxidises the non-metallic impurities to acidic oxides, which escape as gases or combine with the basic lining of the converter to form slag. Other elements are then added to give the type of steel required.

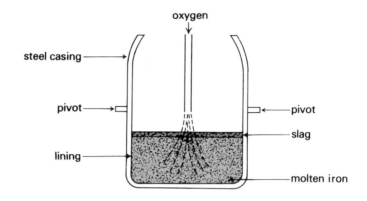

Fig. 18.1

18.3 Manufacture of Sodium Hydroxide and Chlorine

These two important chemicals are manufactured by the electrolysis of brine (sodium chloride solution) in a mercury cell (see Example 18.4).

18.4 Manufacture of Sulphuric Acid

Sulphuric acid is manufactured from sulphur, air and water in the contact process (see Example 18.5).

18.5 Manufacture of Ammonia

Nitrogen from the air and hydrogen from water and natural gas (methane) are combined under special conditions in the Haber process to make ammonia (see Example 18.6).

18.6 Manufacture of Nitric Acid

Nitric acid is made in the Ostwald process, where ammonia is catalytically oxidised by the oxygen of the air (see Example 18.7).

18.7 Worked Examples

Example 18.1

Which of the metals listed below is most likely to be extracted by electrolysis of its molten chloride?
A Calcium
B Copper
C Iron
D Silver
E Zinc

(AEB, 1982)

The answer is **A.**

172

Example 18.2

Figure 18.2 represents a modern aluminium smelter.

The United Kingdom has no natural supply of the ore (empirical formula $Al_2O_3.2H_2O$) from which aluminium is normally extracted. At one time the entire treatment and extraction

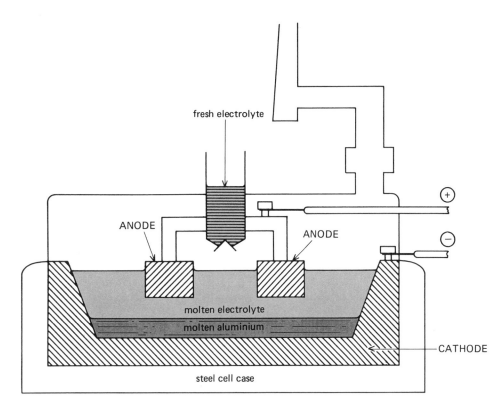

Fig. 18.2

processes were carried out after the importation of the ore, but nowadays it is at least calcined (heated strongly) before shipment.

(a) What is the common name for the ore? **(1 mark)**

Bauxite

(b) What is the effect of *calcining* the ore? **(1 mark)**

It drives off the water from the $Al_2O_3.2H_2O$, leaving Al_2O_3.

(c) What economic advantage is there (i) for the exporter, (ii) for the importer in calcining the ore at source? **(1 mark each)**

(i) The exporter can charge more per tonne for the ore as there is no water present.

(ii) The importer has less to do before extracting the aluminium, thus saving processing costs.

(d) The refined ore (pure aluminium oxide) is dissolved in molten electrolyte in the smelter.

(i) Name the electrolyte used as solvent. **(1 mark)**

Cryolite.

The formula for cryolite is Na_3AlF_6

(ii) Why is the refined ore not used on its own? **(1 mark)**

Electrolysis occurs only in liquids, not in solids. Aluminium oxide has too high a melting point to be used on its own, so it is dissolved in molten cryolite at about 900°C.

(e) Describe the chief pollutant produced by an aluminium smelter. **(2 marks)**

The chief pollutant is gaseous carbon dioxide, formed at the anodes as described in (f) (ii).

(f) (i) Of what material are the electrodes constructed? **(1 mark)**

Carbon.

(ii) Why must the anodes be constantly replaced? **(2 marks)**

Oxygen is produced by the discharge of oxide ions at the anodes.

$$2O^{2-}(l) - 4e^- \rightarrow O_2(g)$$

The oxygen oxidises the anodes to carbon dioxide and so they have to be replaced.

(g) Give the equation for the reaction that occurs at the cathode during electrolysis. **(2 marks)**

$$Al^{3+}(l) + 3e^- \rightarrow Al(l)$$

(h) Why is aluminium usually used in the form of an alloy rather than as a pure metal? **(1 mark)**

Pure aluminium is soft. It is alloyed with other metals to give it more strength.

(i) What are the products of the reaction between aluminium and aqueous sodium hydroxide? **(2 marks)**

(AEB, 1981)

Sodium aluminate and hydrogen are produced.

See Section 12.3.

Example 18.3

Figure 18.3 shows a blast furnace for producing iron.

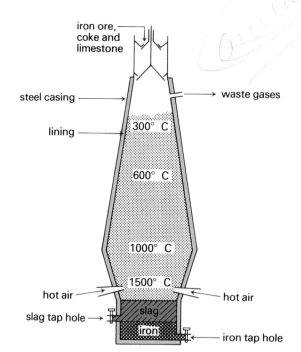

Fig. 18.3

(a) Name the starting materials used in the process. **(4 marks)**

Air, coke, limestone and iron ore (e.g. haematite, Fe$_2$O$_3$).

(b) Why is there a double cup-and-cone device at the top of the furnace? **(1 mark)**

To prevent the escape of dust and gas, which would pollute the atmosphere.

(c) What reaction takes place at the base of the furnace? **(2 marks)**

The coke burns in the hot air blast to form carbon dioxide.

$C(s) + O_2(g) \rightarrow CO_2(g)$

(d) Give an equation for the reduction of the product of (c) by the hot carbon higher up the furnace. **(1 mark)**

$C(s) + CO_2(g) \rightarrow 2CO(g)$

(e) How is the gas which is piped off at the top of the furnace used to keep down costs? **(1 mark)**

It is burned to heat the air for the hot-air blast.

(f) Write an equation for the thermal decomposition of the limestone. **(1 mark)**

$CaCO_3(s) \rightarrow CaO(s) + CO_2(g)$

(g) How is the slag formed? Why is it made? **(3 marks)**

The calcium oxide (basic) formed in (f) reacts with the acidic sandy impurities in the ore (mainly silicon(IV) oxide) to form slag. Slag has a lower melting point than sand and melts, separating from the iron instead of contaminating it.

(h) Give two uses for the slag. **(2 marks)**

Road making, lightweight building materials.

(i) Why is it important to find uses for the slag? **(1 mark)**

To prevent the formation of unsightly slag heaps.

Example 18.4

(a) Describe how sodium chloride is extracted from underground deposits other than by digging it out.

   Discuss the environmental problem caused by the method of extraction you have described, and explain why it is difficult to control. **(5 marks)**

Water is pumped underground, the sodium chloride dissolves and the resulting solution is pumped to the surface.

Underground caverns are formed, and collapse of these can cause the land to sink. It is difficult to fill the caverns with other material, and therefore difficult to prevent the land from sinking. Serious damage to buildings can result from this.

(b) In more than one region the occurrence of sizeable workable deposits of sodium chloride has led to the establishment of large chemical industries in the area. These industries depend on sodium chloride as raw material.

   The scheme below represents a small part only of such a development.

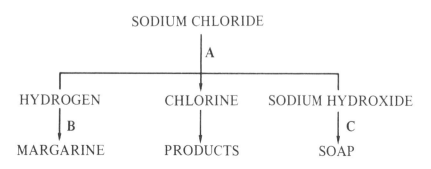

175

(i) Describe the conversions **A**, **B** and **C**.   **(13 marks)**

Conversion A is carried out in the mercury cell (Fig. 18.4).

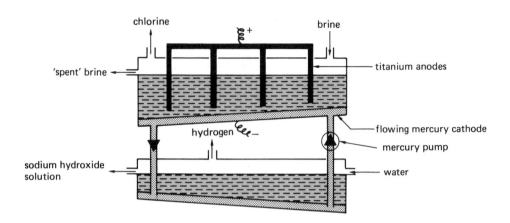

Fig. 18.4

Chlorine is liberated at titanium anodes and is piped off as shown. Sodium is produced at the mercury cathode.

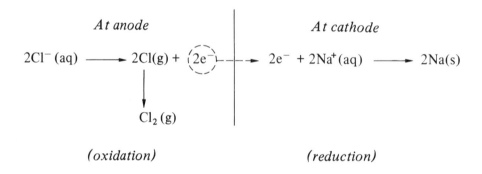

The sodium dissolves in the mercury, and the resulting sodium amalgam drops into water where it reacts to form sodium hydroxide solution and hydrogen, the mercury being regenerated.

$2Na/Hg(l) + 2H_2O(l) \rightarrow 2NaOH(aq) + H_2(g)$

The mercury is pumped back to the top cell and used again. The sodium hydroxide solution is evaporated to dryness and some of the hydrogen is burned in some of the chlorine to make hydrogen chloride.

NB The use of a mercury cathode results in the discharge of sodium ions, not hydrogen ions as would be expected from the relative positions of the two elements in the reactivity series. Careful control of the process is necessary in order to prevent escape of mercury, which has caused serious health problems in the past.

Conversion B is carried out by reacting vegetable oils, which are unsaturated compounds containing double bonds (see Section 19.3), with hydrogen in the presence of a finely divided nickel catalyst at 200°C. An addition reaction takes place.

$$\begin{array}{ccc} \text{H} & \text{H} & \\ | & | & \\ \cdots -\text{C}=\text{C}- \cdots \ + \ \text{H}_2 & \longrightarrow & \cdots -\text{C}-\text{C}- \cdots \\ & & | \quad | \\ & & \text{H} \quad \text{H} \end{array}$$

Conversion C is described in Section 19.7.

    (ii) Name **one** product that is manufactured from chlorine and give **one** use of it.

(2 marks)

Polyvinyl chloride (PVC) is made from chlorine. It is used for guttering and drain-pipes.

Example 18.5

    (a) Outline briefly the manufacture of sulphuric acid starting from sulphur and air (details of industrial plant are **not** required).   **(6 marks)**

Sulphur is burned in excess *air to form sulphur dioxide.*

$S(s) + O_2(g) \rightarrow SO_2(g)$

The gases are cleaned to remove catalyst poisons, and dried by passing them through concentrated sulphuric acid. They are then heated in a heat exchanger by hot gases leaving the catalyst chamber.

In the catalyst chamber sulphur dioxide combines with oxygen from the air in the presence of a complex catalyst based on vanadium(V) oxide. The catalyst is spread out on silica gel to give a greater surface area on which reaction can take place.

$2SO_2(g) + O_2(g) \rightleftharpoons 2SO_3(g)$

The reaction is exothermic and the rate of flow of gases is adjusted to keep the temperature at 450°C without external heating. A pressure slightly above atmospheric is used to force the gases through the plant.

The hot sulphur trioxide is passed back to the heat exchanger to heat the incoming gases and then dissolved in concentrated sulphuric acid. It is not dissolved in water because the reaction is too violent and inefficient. The resulting product, fuming sulphuric acid (oleum), is then diluted with water to give the ordinary concentrated acid.

$H_2SO_4(l) + SO_3(g) \rightarrow H_2S_2O_7(l)$ *(oleum)*
$H_2S_2O_7(l) + H_2O(l) \rightarrow 2H_2SO_4(l)$

NB 1. Care must be taken to minimise the escape of sulphur dioxide into the atmosphere.
    2. The process produces more heat energy than it absorbs, and this, when properly used, helps to improve the cost efficiency of the plant.

    (b) Explain what is observed if
        (i) concentrated sulphuric acid is heated with hydrated copper(II) sulphate crystals.

(3 marks)

The blue hydrated crystals are dehydrated by the concentrated sulphuric acid to give white anhydrous copper(II) sulphate.

        (ii) dilute sulphuric acid is heated with small lumps of calcium carbonate.   **(3 marks)**

Very little reaction occurs because the lumps of calcium carbonate become coated almost immediately with insoluble calcium sulphate.

$CaCO_3(s) + H_2SO_4(aq) \rightarrow CaSO_4(s) + H_2O(l) + CO_2(g)$

(c) What mass of pure sulphuric acid is needed to prepare 1 dm^3 of 2M sulphuric acid? This diluted acid (i.e. 1 dm^3 of 2M strength) is treated with 100 g of magnesium metal. How much metal is left at the end of the reaction?   **(8 marks)**

(OLE)

1 dm^3 of 2M sulphuric acid contains 2 mol of H_2SO_4 *(see Section 20.3)*
i.e. 2 × 98 = 196 g of H_2SO_4
$Mg(s) + H_2SO_4(aq) \rightarrow MgSO_4(aq) + H_2(g)$
mol H_2SO_4 : *mol* Mg = 1 : 1 . . . . A
mol H_2SO_4 = 2
∴ *mol* Mg *used* = 2 *(from A)*
∴   *g* Mg *used* = 2 × 24 = 48
∴   *g* Mg *left* = (100 − 48)
= 52

Example 18.6

(a) Ammonia is manufactured from hydrogen and nitrogen by the Haber process.
(i) What are the raw materials for the process?   **(3 marks)**
Air, water and natural gas (methane).

(ii) Write an equation for the synthesis of ammonia, and state the three essential conditions for an acceptable yield of ammonia.   **(3 marks)**
$N_2(g) + 3H_2(g) \rightleftharpoons 2NH_3(g)$
The plant operates at about 500°C and 250 atm pressure.
A catalyst of iron is used to speed up the reaction.

(b) (i) Describe, with a diagram, how ammonia may be safely dissolved in water in the laboratory.   **(3 marks)**

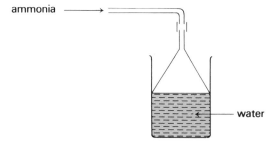

Fig. 18.5

Ammonia is so soluble in water that precautions have to be taken to prevent sucking back. In the funnel arrangement water rises in the funnel but the level in the beaker drops. Air enters under the rim of the funnel and equalises the pressure inside and outside the apparatus so that the water drops back again. The funnel must be almost as wide as the beaker if the system is to work efficiently.

(ii) Aqueous ammonia (ammonium hydroxide solution) is a *weak electrolyte*, while sodium hydroxide solution is a *strong electrolyte*. Explain the meaning of the statement.   **(2 marks)**
A weak electrolyte is a poor conductor of electricity. Aqueous ammonia contains mainly unionised water and ammonia molecules with just a few ions to carry an electric current.

$NH_3(aq) + H_2O(l) \rightleftharpoons NH_4^+(aq) + OH^-(aq)$

Sodium hydroxide is an ionic solid, made up of Na$^+$ *and* OH$^-$ *ions. When it dissolves in water, the ions separate and their large numbers mean that the solution conducts electricity well, i.e. is a strong electrolyte.*

    (c) (i) The fertiliser 'Growmore' contains nitrogen in the form of ammonium compounds and not in the form of nitrates. How would you prove the truth of this statement?

(4 marks)

Carry out tests for ammonium ions and nitrate ions as described in Section 20.2. The first should be positive and the second negative.

        (ii) To what general class of compounds are nitrates converted in plants?   **(2 marks)**

Nitrates are converted to proteins in plants.

Example 18.7

Nitric acid, HNO_3, is made from ammonia in three stages.

Stage 1 A mixture of ammonia and air is passed over a platinum gauze catalyst when nitrogen oxide is formed by an exothermic reaction.

$$4NH_3(g) + 5O_2(g) \rightleftharpoons 6H_2O(g) + 4NO(g)$$

Stage 2 The nitrogen oxide is cooled and mixed with air to give nitrogen dioxide.

$$4NO(g) + 2O_2(g) \rightarrow 4NO_2(g)$$

Stage 3 The nitrogen dioxide forms nitric acid when dissolved in water in the presence of air.

$$4NO_2(g) + 2H_2O(l) + O_2(g) \rightarrow 4HNO_3(aq)$$

    (a) Air is used to provide the necessary oxygen for the process. Give the names of TWO other gases which must inevitably be present in the reaction vessel from this source.

(1 mark)

Nitrogen, carbon dioxide

    (b) What percentage of the air is oxygen?   **(1 mark)**

21%

See Section 10.1.

    (c) Suggest ONE reason why the platinum catalyst used is in the form of a gauze. **(1 mark)**

The gauze has a greater surface area than a lump of platinum of the same mass and is thus more effective.

    (d) When the reaction vessel is first being brought into service, the catalyst is heated electrically. Explain why this heating is stopped once the reactor is in full operation.

(1 mark)

The chemical reaction in Stage 1 takes place on the catalyst surface and is exothermic, so it keeps the catalyst hot.

    (e) Why is the temperature used in Stage 1 reasonably *high*?   **(1 mark)**

So that the reaction proceeds at a reasonable rate.

    (f) What volume of ammonia, measured at room temperature and atmospheric pressure, is needed to produce 63 kg of nitric acid? (1 mol of molecules of any gas occupies 24 dm^3 at room temperature and atmospheric pressure. Relative atomic masses: H = 1, N = 14, O = 16)   **(2 marks)**

1 mol of nitric acid is made from 1 mol of ammonia (see equations)

$\therefore$ *63 g of nitric acid is made from 24 dm^3 of ammonia at room temperature and pressure*

$\therefore$ *63 kg of nitric acid is made from 24 000 dm^3 of ammonia at room temperature and pressure.*

(g) A pressure of up to 10 atmospheres is often used in Stage 1. Suggest a reason for compressing the gases. **(1 mark)**

Compressing the gases increases their concentrations and therefore increases the rate of reaction.

(h) State ONE large-scale use of nitric acid. **(1 mark)**

To make fertilisers such as ammonium nitrate.

(i) Complete the following equation to show how pure nitric acid reacts with water to produce an acidic solution. **(1 mark)**

$HNO_3 + H_2O \rightarrow$

$HNO_3 + H_2O \rightarrow H_3O^+ + NO_3^-$

(j) Electrolysis of a dilute solution of nitric acid produces gases at *both* electrodes. Which gas is likely to be produced at the cathode if inert electrodes are used? **(1 mark)**

(L)

Hydrogen

H_3O^+ ions give hydrogen and water when discharged.

18.8 Self-test Questions

Question 18.1

In the manufacture of aluminium, the reaction occurring at the cathode of the cell is:

A $Al^+ + e^- \rightarrow Al$

B $Al^{3+} - 3e^- \rightarrow Al$

C $Al^{3+} + 3e^- \rightarrow Al$

D $Al \rightarrow Al^{3+} + 3e^-$

E $Al + 3e^- \rightarrow Al^{3+}$

Question 18.2

The production of iron in the blast furnace requires an iron ore and

A coke, limestone and calcium silicate only

B air, limestone and calcium silicate only

C air, coke and limestone only

D air, coke and calcium silicate only

E air, coke, limestone and calcium silicate

(NISEC)

Question 18.3

Discuss the following statement.

The Haber process was developed to make ammonia from nitrogen in the air because nitrogen occurs only in the air.

18.9   Answers to Self-test Questions

18.1 **C**.

18.2 **C**.

18.3 Although the first part of the sentence is true, the second is not.
Nitrogen occurs in ammonium salts, nitrates and many other compounds.

19 Organic Chemistry

19.1 Introduction

Organic chemistry is the chemistry of carbon compounds, excluding carbonates and the oxides of carbon. Many of these compounds are associated with the living world, e.g. proteins and sugars are organic compounds. There are a very large number of organic compounds since carbon atoms can join together in many different ways to form rings and chains of atoms. These compounds can be divided into families known as **homologous** series, in which all the members can be represented by the same general formula. The members have similar chemical properties and the physical properties change gradually as the relative molecular mass changes (see Example 19.2).

Isomerism is the existence of two or more compounds with the same molecular formula but different structures, the different forms being called *isomers* (see Question 19.1). For example, butane and methylpropane, whose structures are given below, are isomers.

butane   methylpropane

19.2 The Alkanes

The alkanes are hydrocarbons, i.e. compounds containing hydrogen and carbon only. They form a homologous series of general formula, $C_n H_{2n+2}$, their molecules containing only single covalent bonds.
Examples include

methane   ethane   propane   butane

Compounds which have molecules containing only single covalent bonds are said to be **saturated**.

Alkanes are obtained commercially from petroleum.

The simplest alkanes are colourless gases or liquids which are insoluble in water but readily dissolve in organic solvents. Like all hydrocarbons, alkanes burn in a plentiful air supply to form carbon dioxide and water. In a limited air supply, carbon monoxide or even carbon may be produced because of incomplete combustion. Alkanes take part in *substitution* reactions, e.g. methane reacts with chlorine or bromine in the presence of light.

$$CH_4(g) + Cl_2(g) \rightarrow HCl(g) + CH_3Cl(g)$$
$$\text{chloromethane}$$

Further substitution may occur to give a mixture of products (see Example 19.2).

Alkanes are used as fuels and as the starting materials in the manufacture of a number of other substances, e.g. plastics and soapless detergents.

19.3 The Alkenes

The alkenes form a homologous series of general formula C_nH_{2n}, their molecules containing a double bond, e.g.

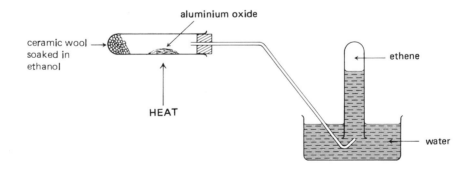

ethene   propene

Compounds which have molecules containing double or triple covalent bonds are said to be **unsaturated**.

Ethene can be prepared by the catalytic dehydration of ethanol in the apparatus shown in Fig. 19.1. The catalyst of aluminium oxide is heated strongly and sufficient heat is conducted down the tube to vaporise the ethanol.

```
                           aluminium oxide

ceramic wool ─→  ▓▓▓   ░░░  ▨▨                          ┌─┐ ─── ethene
soaked in        ▓▓▓   ░░░  ▨▨ ────────┐                │ │
ethanol                      ↑          │              ╱ ▒▒ ╲
                                        │            ╱ ▒▒▒▒▒ ╲
                     ↑                  └──────────────▒▒▒▒▒▒──── water
                   HEAT                             ╲▒▒▒▒▒▒▒╱
                                                     ▒▒▒▒▒
```

Fig. 19.1

$$C_2H_5OH(g) \rightarrow C_2H_4(g) + H_2O(g)$$

Ethene is a colourless gas which is practically insoluble in water but readily soluble in organic solvents.

One of the two bonds in a double bond is easily broken and as a result alkenes are very reactive. They take part in *addition* reactions (see Examples 19.2 and 19.5). When ethene is shaken with a solution of bromine in 1,1,1-trichloroethane, the reddish-brown colour of the bromine disappears almost immediately and 1,2-dibromoethane is obtained as the organic product.

Polymerisation is the process in which many small molecules join together to make one large one,
e.g.

$$nC_2H_4 \longrightarrow (C_2H_4)n$$

Molecules of ethene (the monomer) join together in chains to form polythene (the polymer). One of the bonds in the C=C bond breaks and the molecules undergo an addition reaction. This process is called *addition polymerisation* (see Example 19.3).

Plastics can be divided into two main classes according to their behaviour when heated — thermoplastics or thermosetting plastics (see Example 19.4).

Alkenes are converted into a wide range of products, e.g. plastics, ethanol, 1,2-dibromoethane (a petrol additive) and ethane-1,2-diol (antifreeze).

19.4 Petroleum and Coal

Petroleum was produced underground by the combined effects of heat, pressure and bacteria on the remains of marine animals and plants which died many millions of years ago. It is composed chiefly of hydrocarbons and is the source of most organic compounds used in manufacturing processes. Liquid petroleum is fractionally distilled in tall fractionating towers. Each fraction contains groups of hydrocarbons with boiling points within a particular range, rather than pure samples of individual compounds.

Table 19.1 Crude petroleum fractions

Name of fraction	Number of carbon atoms in fraction	Boiling point	Flammability
Gases			
Petrol			
Naphtha			
Kerosene	increasing	increasing	increasing
Diesel oil			
Lubricating oil			
Bitumen residue			

There is little demand for the heavier oils, and they are usually converted to more useful products by cracking (see Example 19.6).

Cracking is the thermal decomposition of alkanes of high relative molecular mass to give a mixture of alkanes and alkenes of lower relative molecular mass.

Coal was formed in a similar way to petroleum by the underground compression of vegetable matter. It consists mainly of carbon, together with compounds containing hydrogen, oxygen, nitrogen and sulphur. Burning coal pollutes the atmosphere with smoke and harmful gases such as sulphur dioxide. However, *destructive distillation* of coal (heating coal in the absence of air) gives useful products from which many organic chemicals can be obtained.

Petroleum and coal are non-renewable energy resources, i.e. they cannot be replaced. At the present rate, coal could last for at least 200 years but our oil reserves could dry up within 30 years.

19.5 The Alcohols

The alcohols form a homologous series of general formula $C_nH_{2n+1}OH$, e.g.

methanol   ethanol

Ethanol is manufactured by the hydration of ethene.

$$C_2H_4(g) + H_2O(g) \xrightarrow[\text{phosphoric(V) acid catalyst}]{300°C, 65 \text{ atm,}} CH_3CH_2OH(g)$$

It is also prepared by fermentation of aqueous glucose solutions using yeast which contains the enzyme zymase. Enzymes are complex organic catalysts. After filtering, the solution can be concentrated using fractional distillation (see Example 19.8).

$$C_6H_{12}O_6(aq) \rightarrow 2C_2H_5OH(aq) + 2CO_2(g)$$

Fermentation is the slow decomposition of an organic substance brought about by micro-organisms and usually accompanied by the evolution of heat and a gas.

The simple alcohols are colourless liquids which are miscible with water.

(a) Chemical Properties of the Alcohols

1. The alcohols burn in air to form carbon dioxide and water.

$$C_2H_5OH(l) + 3O_2(g) \rightarrow 2CO_2(g) + 3H_2O(g)$$

2. Alcohols are oxidised to acids with acidified potassium dichromate(VI) (see Example 19.8).

185

(b) **Uses of the Alcohols**

1. As solvents.
2. As fuels.
3. In beers, wines and spirits.

Beer is made by the fermentation of the starch in barley; wine is made in a similar way by the fermentation of the sugars in grapes. Distillation of these dilute solutions produced by fermentation increases the alcohol content and yields spirits. During fermentation it is important to exclude air because oxidising bacteria may enter and convert the ethanol to ethanoic acid (vinegar). This accounts for the 'souring' of wine exposed to the air.

19.6 Ethanoic Acid

Ethanoic acid has the formula

$$H-\overset{\overset{\displaystyle H}{|}}{\underset{\underset{\displaystyle H}{|}}{C}}-C\overset{\displaystyle O}{\underset{\displaystyle O-H}{<}}$$

It is prepared by the oxidation of ethanol using acidified potassium dichromate(VI) solution (see Example 19.8).

$$CH_3CH_2OH(l) \xrightarrow{2[O]} CH_3COOH(l) + H_2O(l)$$

Ethanoic acid is a colourless liquid which is completely miscible with water. A 5% solution of ethanoic acid is used as vinegar. It is a weak acid, showing all the typical properties of an acid (see Section 12.2).

19.7 Esters and Detergents

Organic acids react reversibly with alcohols to form esters and water, the process being known as *esterification*.

$$CH_3COOH(l) + C_2H_5OH(l) \rightleftharpoons CH_3COOC_2H_5(l) + H_2O(l)$$
ethanoic acid   ethanol   ethyl ethanoate   water

The reaction is speeded up by heat and by the addition of a catalyst of concentrated sulphuric acid.

The simple esters are colourless liquids which are immiscible with water and have characteristic pleasant smells. They are hydrolysed by warming with dilute sulphuric acid to give an acid and an alcohol (i.e. the reverse of esterification). Sodium hydroxide solution may be used to catalyse the hydrolysis but in this case the reaction goes to completion.

$$CH_3COOC_2H_5(l) + NaOH(aq) \rightarrow CH_3COONa(aq) + C_2H_5OH(l)$$

Soaps are the sodium or potassium salts of long-chained carboxylic acids (e.g. octadecanoic acid) which are prepared by boiling animal fats or vegetable oils (which are esters) with sodium hydroxide solution (see Example 7.4). This process is called saponification. Soapless detergents are made from petroleum and consist of molecules which are similar in form and action to those of soap. Unlike soap, their calcium and magnesium salts are soluble in water and thus do not form scum (see Example 19.9).

19.8 Worked Examples

Example 19.1

(a) (i) What is understood by the term *unsaturated hydrocarbon*? **(2 marks)**

Hydrocarbons are compounds containing carbon and hydrogen only. If the compound is unsaturated, its molecules contain double or triple covalent bonds.

(ii) Name **one** unsaturated hydrocarbon and draw its structural formula. **(2 marks)**

Ethene

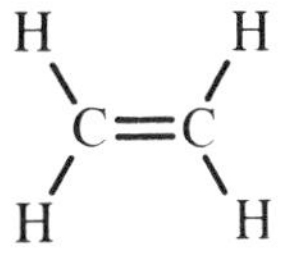

(iii) Name the product formed when bromine and the unsaturated hydrocarbon combine, and write an equation for the reaction. **(3 marks)**

1, 2-dibromoethane

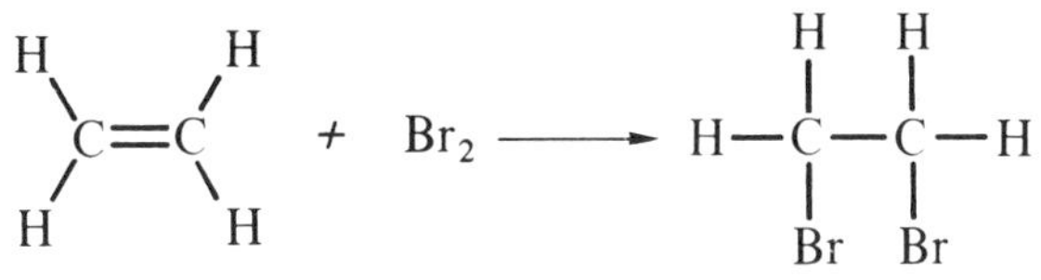

i.e. $C_2H_4 + Br_2 \longrightarrow CH_2BrCH_2Br$

In this reaction two bromine atoms add on across the double bond.

(iv) Name **one** *saturated* hydrocarbon and draw its structural formula. **(2 marks)**

Ethane.

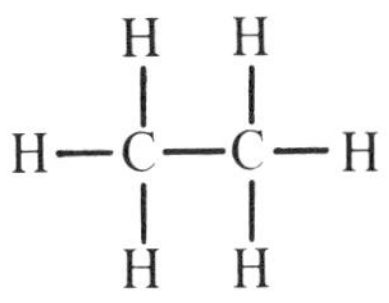

(v) What would be the products of the reaction between 1 mol of this hydrocarbon and 1 mol of bromine? Write an equation for the reaction. **(3 marks)**

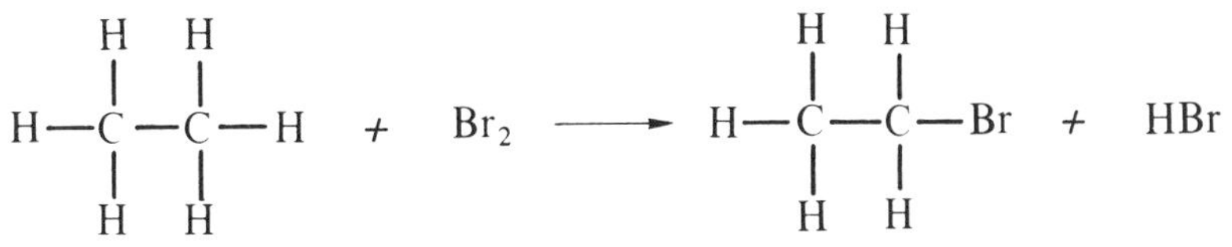

i.e. $C_2H_6 + Br_2 \longrightarrow C_2H_5Br + HBr$

One of the hydrogen atoms in ethane is substituted by a bromine atom to give bromoethane and hydrogen bromide.

In both (iii) and (v) the colour of the bromine disappears and colourless products are obtained, but (iii) is much quicker than (v).

(b) (i) Which of the two named hydrocarbons can be converted to a polymer? Name the polymer and draw its structure. **(3 marks)**

Unsaturated hydrocarbons can be converted to polymers, i.e. ethene can be polymerised to give polythene,

$$\cdots -\underset{\underset{H}{|}}{\overset{\overset{H}{|}}{C}}-\underset{\underset{H}{|}}{\overset{\overset{H}{|}}{C}}-\underset{\underset{H}{|}}{\overset{\overset{H}{|}}{C}}-\underset{\underset{H}{|}}{\overset{\overset{H}{|}}{C}}- \cdots$$

(ii) Explain what is meant by polymerisation. **(2 marks)**

Polymerisation is the process in which many small molecules join together to make one large one. One of the bonds in each carbon-to-carbon double bond breaks and then the molecules add together to make the polymer.

(iii) State one environmental disadvantage of a polymer such as polythene. **(1 mark)**

It does not rot away naturally in the environment, i.e. it is non-biodegradable.

(iv) Would burning be a suitable way of disposing of a plastic such as PVC? Explain your answer. **(2 marks)**

No. On burning, PVC releases hydrogen chloride. This has a choking smell and dissolves in rain water to give hydrochloric acid, another component of acid rain (see Example 16.4).

Example 19.2

(a) Methane is the first member of the *homologous series* of hydrocarbons called the *alkanes*. Explain what is meant by the terms printed in italics. **(6 marks, 2 marks)**

A homologous series is a group of compounds in which

1. all members possess the same general formula,

2. each member differs from the next by the group $-CH_2-$,

3. the physical properties of members change gradually as the relative molecular mass changes,

4. all members have similar chemical properties, and

5. all members can be made by general methods.

The alkanes form a homologous series of general formula C_nH_{2n+2}.

(b) The alkenes form a homologous series of general formula C_nH_{2n}. Write down the molecular formula of the third member. **(2 marks)**

C_4H_8.

Ethene, C_2H_4, is the first member.

(c) Write an equation for, and name the products of the reaction between
(i) chlorine and ethane,

$C_2H_6(g) + Cl_2(g) \longrightarrow HCl(g) + C_2H_5Cl(g)$

hydrogen chloride   *chloroethane*

The reaction does not stop here but successive hydrogen atoms in the chloroethane will be replaced by chlorine atoms to give eventually hexachloroethane

$$Cl-\underset{\underset{\displaystyle Cl}{|}}{\overset{\overset{\displaystyle Cl}{|}}{C}}-\underset{\underset{\displaystyle Cl}{|}}{\overset{\overset{\displaystyle Cl}{|}}{C}}-Cl$$

(ii) chlorine and ethene. **(6 marks)**

$$\underset{H}{\overset{H}{>}}C=C\underset{H}{\overset{H}{<}} \; + \; Cl_2 \longrightarrow H-\underset{\underset{\displaystyle Cl}{|}}{\overset{\overset{\displaystyle H}{|}}{C}}-\underset{\underset{\displaystyle Cl}{|}}{\overset{\overset{\displaystyle H}{|}}{C}}-H$$

1,2-dichloroethane

(d) Use the reactions in (c) to explain what is meant by a *substitution* and an *addition* reaction. **(4 marks)**
(OLE)

The reaction between ethane and chlorine is an example of a substitution reaction.

A substitution reaction is a reaction in which one atom or group of atoms in a molecule is replaced by another. In this example, successive hydrogen atoms are replaced by chlorine atoms.

The reaction between ethene and chlorine is an example of an addition reaction.

An addition reaction is a reaction in which two or more molecules react to give a single product.

Example 19.3

Buta-1,3-diene forms a polymer according to the equation

$$n\, CH_2 = CH - CH = CH_2 \longrightarrow \{CH_2 - CH = CH - CH_2\}_n$$

(a) Explain this reaction. **(3 marks)**

This is an addition polymerisation process. In this process, the molecules of the monomer (buta-1,3-diene) join together to form the polymer. The empirical formula of the polymer is the same as that of the monomer.

(b) What does the n on the left-hand side of the equation stand for? **(1 mark)**
n is a large whole number, usually several hundred.

(c) Give one chemical test which would show that the product still contains double bonds. **(3 marks)**
(OLE)

A solution of bromine in 1,1,1-trichloroethane is rapidly decolorised.

189

Example 19.4

(a) Rewrite in a balanced form the equation given below. **(2 marks)**

$$CH_2 = CH_2 + O_2 + HCl \rightarrow Cl-CH = CH_2 + H_2O$$
chloroethene

$$2CH_2 = CH_2 + O_2 + 2HCl \rightarrow 2Cl - CH = CH_2 + 2H_2O$$

(b) Show by means of an equation how chloroethene may be converted into a polymeric product. **(2 marks)**

poly(chloroethene),
polyvinylchloride, PVC

(c) How could it be demonstrated that the polymer is a thermoplastic rather than a thermo-setting polymer? **(2 marks)**

Thermoplastics are those that soften on heating and harden on cooling, the process being repeatable any number of times. Thermosetting plastics soften on being heated the first time after manufacture and thus can be moulded. This heating causes cross-linking to occur between the long-chained macromolecules, and results in the setting up of a rigid three-dimensional network which cannot be softened by subsequent reheating. Thus, studying the effect of heat on PVC will demonstrate that it is a thermoplastic rather than a thermosetting polymer.

(d) Calculate the volume of chloroethene (measured at room temperature and pressure) which would be necessary to produce 125 g of polymer, assuming 100% yield. **(4 marks)**

Relative molecular mass of polymer = n(3 + 24 + 35.5)
= n × 62.5
2 mol of chloroethene are needed to produce 125 g of polymer
Volume of chloroethene = 2 × 24 dm³ (molar volume = 24 dm³ mol⁻¹ at room temperature and pressure)
= 48 dm³

Example 19.5

Give the name or molecular formula for the product of reaction of ethene (ethylene) with each of the following:

(i) bromine; **(1 mark)**

$C_2H_4Br_2$ *or 1,2-dibromoethane*

(ii) steam in the presence of a hot phosphoric acid catalyst; **(1 mark)**

C_2H_5OH *or ethanol*

(iii) hydrogen with a suitable catalyst; **(1 mark)**

C_2H_6 *or ethane*

The conditions needed are a nickel catalyst and a temperature of 200°C.
See Example 11.6.

(iv) more ethene. **(1 mark)**
$(C_2H_4)_n$ *or polythene* (OLE)

Example 19.6

Domestic paraffin, which is a mixture of alkanes, is passed over strongly heated broken porcelain in the apparatus shown in Fig. 19.2.

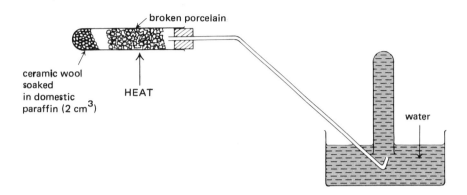

Fig. 19.2

(a) (i) What is the purpose of the broken porcelain? **(1 mark)**
The porcelain enables the paraffin vapour to be heated to a high temperature in order to decompose. It ensures maximum contact between the vapour and the hot surface.

(ii) What is the name given to this type of thermal decomposition? **(1 mark)**
Cracking.

(b) If propane is heated under suitable conditions, it decomposes into methane and ethene.
(i) Write an equation for this reaction. **(1 mark)**
$C_3H_8(g) \rightarrow CH_4(g) + C_2H_4(g)$

N.B. Cracking always produces an unsaturated compound as one of the products.

(ii) How would you prove that ethene is formed? **(2 marks)**
Bromine water is decolorised.

(c) Using different conditions, propane can be decomposed into propene and hydrogen.
(i) Write an equation for this reaction. **(1 mark)**
$C_3H_8(g) \rightarrow C_3H_6(g) + H_2(g)$

(ii) What volume of hydrogen, measured at the same temperature and pressure, is obtained from the decomposition of 100 cm^3 of propane? **(2 marks)**
From the equation
1 mol of propane → 1 mol of hydrogen
1 vol of propane → 1 vol of hydrogen
100 cm^3 of propane → 100 cm^3 of hydrogen

(d) Why are these reactions of importance in the petrochemical industry? **(2 marks)**
There is little demand for the higher alkanes but the demand for the alkanes with fewer numbers of carbon atoms in their molecules is far greater than can be obtained directly from petroleum. The alkanes are converted into simpler alkanes and alkenes by cracking.

Example 19.7

    (a) What is a fossil fuel?   **(2 marks)**
Petroleum and coal are fossil fuels. They are formed from decaying remains of marine animals and plants which died millions of years ago (see Section 19.4).

    (b) Give the name and formula of the main compound found in natural gas.   **(2 marks)**
Name *Methane*   Formula CH_4

    (c) Name two areas where petroleum can be found.   **(2 marks)**
North Sea, Arabian Gulf.

    (d) Petroleum is a mixture of many hydrocarbons. How could you separate kerosene from liquid petroleum? (Table 19.1 may help.)   **(1 mark)**
Fractional distillation.

    (e) Give one use of
        (i) kerosene,
Fuel for jet engines.

        (ii) bitumen.   **(2 marks)**
On roads.

    (f) Suggest one hazard involved in the storage of natural gas or petroleum.   **(2 marks)**
These substances have low boiling points and are highly flammable. They must be kept away from all flames or sparks.

    (g) What does the hazard sign in Fig. 19.3 indicate?   **(1 mark)**

Fig. 19.3

Highly flammable.

    (h) Give two adverse effects of coal mining on the environment.   **(2 marks)**
 (i) *subsidence of the land,*
(ii) *spoil heaps.*

In addition, breathing in coal dust affects miners' lungs.

    (i) What information does the graph in Fig. 19.4 give you?   **(4 marks)**
It tells us that the demand for all forms of energy has increased. Although the demand for coal has increased only slightly, oil and natural gas are supplying a larger proportion of our energy needs. In recent years, there has been a large increase in nuclear power and hydroelectricity and a slowing in the increase in use of oil. This corresponds to the large increase in oil prices in the 1970s.

    (j) Suggest two other energy sources.   **(2 marks)**
Sunlight (by means of solar panels).
Rotting vegetable waste. This produces methane, which is made use of in sewage works where sludge digestion is often used to provide sufficient fuel for the works (see Example 2.7).

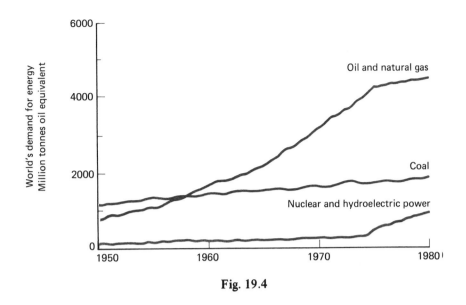

Fig. 19.4

Example 19.8

Fermentation of sucrose by yeast may be carried out in apparatus such as is shown in Fig. 19.5.

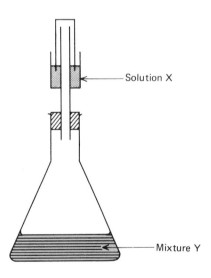

Fig. 19.5

(a) What ingredients are required in mixture **Y** for successful fermentation? Give any other essential conditions. **(4 marks)**

Sugar solution and yeast are required for fermentation. The solution must be kept warm.

(b) A dye, Janus Green, is added to mixture **Y**. It is blue in the presence of dissolved oxygen and pink in its absence. What colour would you expect it to go as fermentation takes place? Explain your answer. **(4 marks)**

Initially I would expect the dye to go blue. As fermentation proceeds, carbon dioxide is produced. This will carry the dissolved oxygen out of solution and hence the dye will gradually go pink.

(c) The gaseous product of fermentation bubbles out through the airlock containing solution **X**. Name this product and say how you would prove its identity using a suitable solution **X**. **(4 marks)**

The gaseous product is carbon dioxide. This can be identified using lime water as solution X. The lime water will turn milky.

(d) (i) The organic product of fermentation may be concentrated by distillation. Name the compound and give its structural formula. **(3 marks)**

Ethanol is obtained by fermentation.

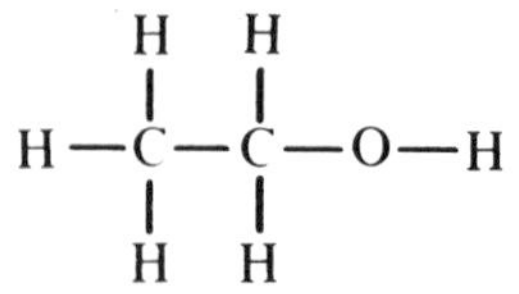

$$H-\overset{\displaystyle H}{\underset{\displaystyle H}{\overset{|}{\underset{|}{C}}}}-\overset{\displaystyle H}{\underset{\displaystyle H}{\overset{|}{\underset{|}{C}}}}-O-H$$

(ii) Describe how you would convert the concentrated product into the corresponding carboxylic acid. **(3 marks)**

(OLE)

Ethanol is heated with acidified potassium dichromate(VI) solution so as to oxidise it to ethanoic acid.

$$CH_3CH_2OH(l) \xrightarrow{\text{oxidation}} CH_3COOH(l) + H_2O(l)$$
$$\text{ethanoic acid}$$

Example 19.9

Below is a simple representation of a detergent molecule.

Covalent tail ————————————● Ionic head

(a) Using representations like the one above, fill in the diagram below to show how you would expect detergent molecules to interact with grease and water. **(2 marks)**

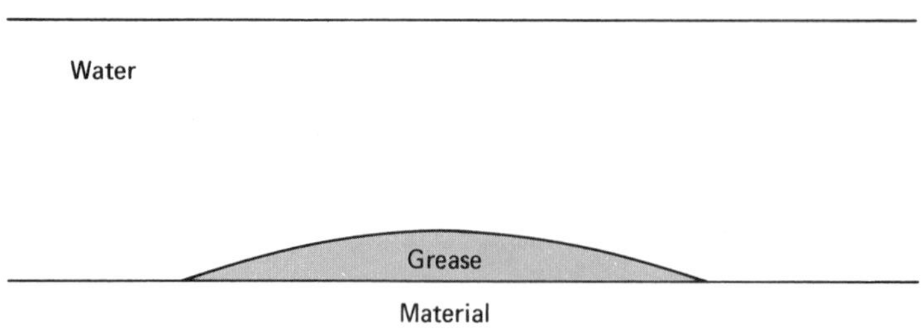

Water

Grease

Material

The hydrophobic (water-hating) tails of the detergent molecules enter the grease but the hydrophilic (water-loving) heads stay in the water. The grease is broken up into small particles and carried off into solution.

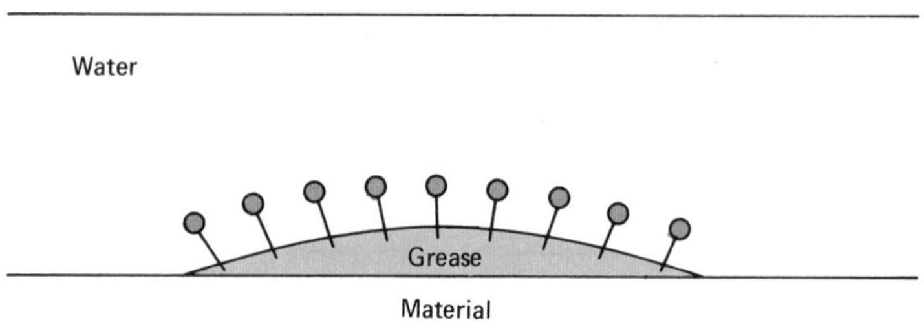

Water

Grease

Material

(b) Describe one similarity in structure between the above detergent molecule and a soap molecule. **(1 mark)**

Detergent molecules and soap molecules are similar in that they both have a long covalent tail and an ionic head.

(c) Most detergents produced today are biodegradable.

(i) Explain what this term means. **(1 mark)**

Biodegradable materials are acted upon by bacteria and decomposed into simpler substances.

(ii) Why is it necessary that detergents should be biodegradable? **(2 marks)**

A large build-up of detergents in our rivers is undesirable since vast quantities of foam are unsightly and marine life is harmed.

(d) (i) What happens when soap is dissolved in water containing calcium ions in solution? **(2 marks)**

The soap reacts with the calcium ions to give a precipitate of calcium octadecanoate (scum).

$$2C_{17}H_{35}COONa(aq) + Ca^{2+}(aq) \rightarrow (C_{17}H_{35}COO)_2Ca(s) + 2Na^+(aq)$$
$$\text{soap} \qquad\qquad\qquad\qquad \text{scum}$$

(ii) Give the name of one calcium compound commonly found in solution in natural waters. **(1 mark)**

(AEB, 1982)

Either calcium sulphate or calcium hydrogencarbonate is a suitable answer.

19.9 Self-test Questions

Question 19.1

(i) Name the straight-chain compound having the formula C_5H_{12}.
Is it an alkane or an alkene? **(2 marks)**
(ii) Draw the structural formulae of the two isomers of this compound. **(2 marks)**

Question 19.2

From the information given below, deduce the identities of the unknown substances, giving a brief explanation in support of your answer, and writing balanced equations for the reactions which take place.

(a) (i) A colourless liquid **A**, molecular formula C_2H_6O, was vaporised and passed over a heated catalyst. A colourless gas **B** was formed which decolorised bromine water when shaken with it.

(ii) Liquid **A** was heated with acidified potassium dichromate(VI) and the product of the reaction was isolated; this product **C** was a sharp-smelling liquid with a molecular formula $C_2H_4O_2$.

(iii) When gas **B** was subjected to a very high pressure and a catalyst, a colourless solid **D** was obtained. **D** was inert to most chemical reagents and had a very high relative molecular mass. **(15 marks)**

(b) An organic compound **E** underwent the following reactions.

(i) It burned completely in air to give carbon dioxide and water only.
(ii) it decolorised bromine water, and
(iii) it had a pH of less than 7.
What can you deduce about the nature of **E**? **(5 marks)**

19.10 Answers to Self-test Questions

19.1 (i) Pentane. An alkane.

(ii) Pentane has the structural formula:

$$H-\underset{\underset{H}{|}}{\overset{\overset{H}{|}}{C}}-\underset{\underset{H}{|}}{\overset{\overset{H}{|}}{C}}-\underset{\underset{H}{|}}{\overset{\overset{H}{|}}{C}}-\underset{\underset{H}{|}}{\overset{\overset{H}{|}}{C}}-\underset{\underset{H}{|}}{\overset{\overset{H}{|}}{C}}-H$$

The question requires you to draw the structural formulae of the other two isomers, which are:

19.2 (a) (i) Since **B** decolorises bromine water, **B** must contain a double bond. **B** is ethene, C_2H_4. **B** is obtained by the dehydration of **A**: **A** is ethanol.

$$C_2H_5OH(g) - H_2O(g) \rightarrow C_2H_4(g)$$
$$\quad\quad \mathbf{A} \quad\quad\quad\quad\quad\quad\quad \mathbf{B}$$
$$C_2H_4(g) + Br_2(aq) \rightarrow C_2H_4Br_2(l)$$

(ii) Alcohols are oxidised to acids on heating with acidified potassium dichromate(VI) solution. Hence **C** is ethanoic acid.

$$C_2H_5OH(l) \xrightarrow{\text{oxidation}} CH_3COOH(l) + H_2O(l)$$

(iii) Gas **B** is ethene, which will polymerise at high pressure in the presence of a catalyst to give **D**, polythene.

$$nC_2H_4(g) \rightarrow (C_2H_4)_n(s)$$

(b) (i) **E** must contain carbon and hydrogen and possibly oxygen.
    (ii) **E** contains a $C = C$ bond.
    (iii) **E** is an acid.
    These three reactions together show that **E** is an organic acid containing a $C = C$ bond.

Something like $CH_3 CH = CHCO_2 H$ is a possibility.

---

20 Chemical Analysis

20.1 Introduction

There are two types of chemical analysis, *qualitative analysis*, in which the various elements in a substance are identified, and *quantitative analysis*, where their relative masses are found.

20.2 Qualitative Analysis

This section has been restricted to the detection of the following ions: NH_4^+, K^+, Na^+, Ca^{2+}, Al^{3+}, Zn^{2+}, Fe^{2+}, Fe^{3+}, Cu^{2+}, CO_3^{2-}, SO_4^{2-}, NO_3^- and Cl^-.

Tests 1–6 are carried out on a small quantity of the *solid* substance. Tests 7 and 8 on a *solution* of the substance in water or acid, and Test 9 on *either* solid or solution.

Experiment	Observation	Conclusion
1 Note appearance of solid	(a) Blue or bright green (b) Pale green (c) Yellow or brown	Possibly Cu^{2+} Possibly Fe^{2+} Possibly Fe^{3+}
2 Heat in an ignition tube	(a) O_2 (glowing splint relit), no brown fumes of NO_2 (b) O_2 (glowing splint relit), + brown fumes of NO_2 (c) CO_2 (lime water milky) (d) Sublimation (e) Steam (f) Yellow when hot, white when cold	Nitrate of K or Na Nitrate (not of K or Na) CO_3^{2-} (or HCO_3^-) Possibly NH_4^+ Water of crystallisation ZnO *residue*
3 Add dil. HCl.	Immediate effervescence (CO_2 – lime water milky)	CO_3^{2-} (or HCO_3^-)
4 Add 2 cm³ conc. H_2SO_4. Warm *gently* if no reaction. Do not pour hot mixture into water.	(a) Same gas as in test 3 (b) HCl (colourless, fuming, acidic gas)	Anions as in 3 Cl^- (Go on to test 5)

5	Dissolve in distilled water (dil. HNO_3 if insoluble in water). Acidify with dil. HNO_3 and add a few drops of $AgNO_3$ solution.	Thick white precipitate (of AgCl), soluble in excess ammonia solution	Cl^-
6	Dissolve in distilled water (dil. HCl if insoluble in water). Acidify with dil. HCl and add a little $BaCl_2$ solution.	Thick white precipitate (of $BaSO_4$)	SO_4^{2-}
7	Add NaOH solution until *just* alkaline (test with indicator paper), then fill up the tube and tip its contents into a second tube. Heat if no precipitate. Add a *little* Al powder if no smell.	(a) White precipitate, insoluble in excess alkali ($Ca(OH)_2$) (b) White precipitate, soluble in excess alkali ($Al(OH)_3$, $Zn(OH)_2$) (c) Pale blue precipitate ($Cu(OH)_2$) (d) Dirty-green precipitate ($Fe(OH)_2$) (e) Red-brown precipitate ($Fe(OH)_3$) (f) NH_3 (smell, alkaline gas) (g) NH_3 as in (f)	Ca^{2+} (Go on to test 9) Al^{3+}, Zn^{2+} Cu^{2+} Fe^{2+} Fe^{3+} NH_4^+ NO_3^-
8	Repeat test 7 using NH_3 solution in place of the NaOH solution.	(a) White precipitate, insoluble in excess alkali ($Ca(OH)_2$, $Al(OH)_3$ (b) White precipitate, soluble in excess alkali ($Zn(OH)_2$) (c) Pale blue precipitate ($Cu(OH)_2$), giving royal blue solution with excess alkali (d) Dirty-green precipitate ($Fe(OH)_2$) (e) Red-brown precipitate ($Fe(OH)_3$)	Ca^{2+}, Al^{3+} Zn^{2+} Cu^{2+} Fe^{2+} Fe^{3+}
9	Dip a platinum wire into conc. HCl in a watch glass and then heat the *end* of the wire in a flame. Repeat if necessary until the wire does not colour the flame. Then dip the wire into the acid again, touch it on the powdered solid and replace it in the flame. Clean the wire after use.	(a) Lilac flame (red through blue glass) (b) Golden yellow flame (c) Brick-red flame (d) Green–blue flame	K^+ Na^+ Ca^{2+} Cu^{2+}

20.3 Quantitative Analysis

In this chapter we shall limit ourselves to the branch of quantitative analysis known as *volumetric analysis*, where a solution of the substance under investigation is titrated with a standard solution of a suitable reagent. (A **standard solution** is one whose concentration is known). Only titrations between acids and alkalis will be considered. (See Examples 20.6 and 20.7, and Question 20.8.)

(a) Concentrations of solutions

Since compounds react in mole ratios, it is usual to quote concentrations of solutions in $mol\ dm^{-3}$. An abbreviation which is often employed for $mol\ dm^{-3}$ is the letter 'M'. Thus a 0.1 M solution of a given substance has a concentration of 0.1 mole of solute per dm^3 of solution. Strictly speaking, this abbreviation is now obsolete.

20.4 Worked Examples

Example 20.1

Two experiments were carried out on a substance X (Fig. 20.1).

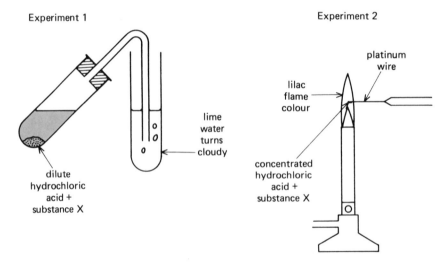

Fig. 20.1

(a) Which gas is turning the lime water (calcium hydroxide solution) cloudy?
Carbon dioxide.

(b) **From experiment 1 alone**, what can be learned about the identity of X?
X is a carbonate (or hydrogencarbonate).

See Test 3.

(c) **From experiment 2 alone**, what can be learned about the identity of X?   **(3 marks)**
(SEB)

X is a compound of potassium.

See Test 9.

Example 20.2

In **each** case below give **ONE** simple test which would enable you to distinguish between the two substances named, stating clearly what would happen when you performed the test on each substance of the pair. (No equations are required.)

(a) Magnesium oxide and calcium oxide. **(3 marks)**

Add a small quantity of water to each solid. Magnesium oxide will not react but calcium oxide will react violently, giving out a great deal of heat and swelling up to give a fine powder of calcium hydroxide.

(b) Calcium hydroxide and calcium carbonate. **(5 marks)**

Add a spatula measure of each solid to dilute hydrochloric acid. Calcium hydroxide will dissolve, forming a solution of calcium chloride and water (acid + base → salt + water). Calcium carbonate will fizz, giving off carbon dioxide (turns lime water milky) and leaving a solution of calcium chloride and water.

See Test 3.

(c) Magnesium sulphate and magnesium chloride. **(5 marks)**

Add dilute hydrochloric acid, followed by barium chloride solution, to an aqueous solution of each substance. Magnesium sulphate will give a white precipitate of barium sulphate but magnesium chloride will not react.

See Test 6.

(d) Zinc hydroxide and magnesium hydroxide. **(5 marks)**
(NISEC)

To each solid add excess sodium hydroxide solution and shake well. Zinc hydroxide is amphoteric (see Section 12.5) and will dissolve, forming a solution of sodium zincate. Magnesium hydroxide will not react.

Example 20.3

In each of the experiments described below, suggest a probable identity for each substance indicated by a letter and write equations for the reactions which take place.

(a) A colourless gas **A** was passed over a heated yellow solid **B**. The products were a shiny solid and a gas **C**, which formed a white precipitate **D** when passed through aqueous calcium hydroxide. **(6 marks)**

C *is carbon dioxide and* **D** *is calcium carbonate.*
$Ca(OH)_2(aq) + CO_2(g) \rightarrow CaCO_3(s) + H_2O(l)$
A *must be oxidised to carbon dioxide and* **B** *must be reduced to a metal. Therefore* **A** *is carbon monoxide and* **B** *is lead(II) oxide.*
$PbO(s) + CO(g) \rightarrow Pb(s) + CO_2(g)$

(b) When excess zinc powder was added to a blue solution **E** and stirred, a red–brown solid **F** was formed and the solution became colourless. The addition of acidified aqueous silver nitrate to the colourless solution resulted in the formation of a white precipitate **G**. **(6 marks)**

G *is silver chloride, so* **E** *must be a chloride of some kind.*

See Test 5.

F is copper, displaced from E by zinc. Therefore E is copper(II) chloride.

$$Zn(s) + CuCl_2(aq) \rightarrow ZnCl_2(aq) + Cu(s)$$
$$2AgNO_3(aq) + ZnCl_2(aq) \rightarrow 2AgCl(s) + Zn(NO_3)_2(aq)$$

Example 20.4

Suggest a name and formula for each of the compounds **A** to **G** below. Write equations for the reactions mentioned. Structural formulae are required for the compounds **F** and **G**.

(a) The green crystalline solid **A** dissolved in water to give a pale green solution. One portion of this solution was treated with aqueous ammonia (ammonium hydroxide solution) and a green precipitate **B** was formed. A second portion of the solution of **A** gave a white precipitate **C** with barium chloride solution that was not soluble in dilute hydrochloric acid. **(8 marks)**

C is barium sulphate, $BaSO_4$, *so* **A** *is a sulphate.*

See Test 6.

B is iron(II) hydroxide, $Fe(OH)_2$.

See Test 8.

So **A** *must be iron(II) sulphate,* $FeSO_4$.
$$FeSO_4(aq) + 2NH_4OH(aq) \rightarrow Fe(OH)_2(s) + (NH_4)_2SO_4(aq)$$
$$FeSO_4(aq) + BaCl_2(aq) \rightarrow BaSO_4(s) + FeCl_2(aq)$$

(b) A white deliquescent solid **D** dissolved in water exothermically to give a strongly alkaline solution. When the solution of **D** was added to copper(II) sulphate solution, a blue precipitate **E** was formed. Substance **D** gave a golden yellow flame colour. **(8 marks)**

D is sodium hydroxide, NaOH, *and* **E** *is copper(II) hydroxide,* $Cu(OH)_2$.

See Tests 7 and 9.

$$CuSO_4(aq) + 2NaOH(aq) \rightarrow Cu(OH)_2(s) + Na_2SO_4(aq)$$

(c) When the colourless flammable liquid **F** is boiled with a concentrated acidified potassium dichromate solution and the resulting mixture is distilled, an acidic liquid **G** with a vinegary smell is produced. **(4 marks)**

F is ethanol, **G** *is ethanoic acid.*

F   G

$$C_2H_5OH \xrightarrow{2[O]} CH_3COOH + H_2O$$

See Section 19.6.

Example 20.5

(a) Describe what you would observe when each of the following pairs of substances is mixed, and write an ionic equation for each reaction.
(i) sodium chloride and silver nitrate solutions;

A white precipitate of silver chloride would form.

See Test 5.

$Ag^+(aq) + Cl^-(aq) \rightarrow AgCl(s)$

(ii) iron(III) chloride and ammonium hydroxide solutions;
A brown precipitate of iron(III) hydroxide would form.

See Test 8.

$Fe^{3+}(aq) + 3OH^-(aq) \rightarrow Fe(OH)_3(s)$

(iii) copper(II) sulphate solution and zinc.   (3 × 3 marks)
A reddish–brown precipitate of copper would form, and the blue colour of the solution would fade.
$Cu^{2+}(aq) + Zn(s) \rightarrow Cu(s) + Zn^{2+}(aq)$

(b) State how you would distinguish between the two substances in each of the following pairs; one test for each pair is sufficient, but you should state how each substance will behave under the test you describe:
(i) chlorine and sulphur dioxide;

Bubble each gas into sodium bromide solution. Chlorine displaces bromine and the solution turns reddish–brown. Sulphur dioxide has no effect.
$Cl_2(g) + 2Br^-(aq) \rightarrow 2Cl^-(aq) + Br_2(aq)$

(ii) potassium carbonate and potassium nitrate.   (2 × 3 marks)
Add dilute hydrochloric acid. Potassium carbonate will give an effervescence of carbon dioxide (turns lime water milky). Potassium nitrate will not react.
$K_2CO_3(s) + 2HCl(aq) \rightarrow 2KCl(aq) + H_2O(l) + CO_2(g)$

Example 20.6

(a) The following instructions refer to an experiment to determine the concentration of dilute hydrochloric acid, using 0.1 M sodium hydroxide solution (containing 0.1 mol dm^{-3}).
Rinse a 25.0 cm^3 pipette with distilled water and then with the sodium hydroxide solution. Rinse a burette, including the jet, with distilled water and then with the dilute hydrochloric acid. Fill the burette with the acid, open the tap and allow the acid to run out until all air bubbles have been expelled from the jet. Wash the flask with distilled water *only*.
Transfer 25.0 cm^3 of the alkali to a conical flask by means of the pipette. Add 2-3 drops of screened methyl orange indicator to the alkali and then read the level of the acid in the burette. Run 1 cm^3 portions of the acid into the alkali, swirling the flask after each addition until the indicator changes from green to red. Note the new burette reading and subtract the initial reading from this to give the volume, to the nearest 1 cm^3, of the acid required to neutralise the alkali. Empty the flask, rinse it with distilled water and repeat the titration, this time adding the acid in a continuous stream until the volume is within 1 cm^3 of the end-point. Carry out the final additions dropwise, with swirling, until the indicator turns buff coloured, showing that the solution is exactly neutral. Perform further accurate titrations until two results within 0.1 cm^3 of one another are obtained.

(i) Why are bubbles expelled from the jet before beginning the experiment? **(2 marks)**

Because they will produce an inaccurate reading of volume if they are expelled during the titration.

(ii) What does the colour change of the indicator from green to red show? **(1 mark)**
It shows that the acid is in excess.

(iii) Why is the first rough titration performed? **(1 mark)**
To save time — the accurate titrations can be performed more quickly if the approximate end-point is known.

(iv) Why is the accurate titration repeated? **(1 mark)**
To check that the result is correct.

(b) In a similar experiment, 25.0 cm^3 of 0.1 M sodium hydroxide solution neutralised 22.0 cm^3 of hydrochloric acid.

$$HCl(aq) + NaOH(aq) \rightarrow NaCl(aq) + H_2O(l)$$

Calculate

(i) the number of moles of sodium hydroxide solution used; **(3 marks)**
mol of NaOH *per dm^3* $= 0.1$

$\therefore$ *mol of* NaOH *per 25.0 cm^3* $= 0.1 \times \dfrac{25}{1000} = 0.0025$

(ii) the number of moles of hydrochloric acid in the 22.0 cm^3 samples; **(4 marks)**
mol NaOH : *mol* HCl $= 1 : 1$ *(from equation)*
$\therefore$ *mol of* HCl *per 22.0 cm^3* $= 0.0025$

(iii) the concentration of the hydrochloric acid in moles per dm^3; **(3 marks)**

mol of HCl *per dm^3* $= 0.0025 \times \dfrac{1000}{22} = 0.114$

(iv) the concentration of the acid in grams per dm^3. **(3 marks)**
g of HCl *per dm^3* $= 0.114 \times 36.5$ *(M$_r$(HCl) = 36.5)*
$= 4.16$

Example 20.7

The equation for the reaction between sodium carbonate and an aqueous solution of sulphuric acid is:

$$Na_2CO_3(aq) + H_2SO_4(aq) \rightarrow Na_2SO_4(aq) + CO_2(g) + H_2O(l)$$

In an experiment it was found that 10 cm^3 of a solution of sulphuric acid reacted completely with 40 cm^3 of 0.025 M sodium carbonate solution. What is the concentration of sulphuric acid in the original solution (in mol dm^{-3})?

A 0.1
B 0.5
C 1.0
D 1.5
E 2.0

(L)

The answer is **A**.

$$\text{mol } Na_2CO_3 : \text{mol } H_2SO_4 \quad = 1 : 1 \ldots \ldots \ldots : A$$

$$\text{mol } Na_2CO_3 \text{ per dm}^3 \quad = 0.025$$

$$\text{mol } Na_2CO_3 \text{ per 40 cm}^3 \quad = 0.025 \times \frac{40}{1000}$$

$$\text{mol } H_2SO_4 \text{ per 10 cm}^3 \quad = 0.025 \times \frac{40}{1000} \times 1 \text{ (from A)}$$

$$\text{mol } H_2SO_4 \text{ per dm}^3 \quad = 0.025 \times \frac{40}{1000} \times 1 \times \frac{1000}{10}$$

$$= 0.1$$

N.B. Always begin line A with the substance whose concentration is *known*.

20.5 Self-test Questions

Question 20.1

Describe briefly a simple chemical or physical test which would distinguish between the ions in each of the following pairs:

(a) $Br^-(aq)$ and $NO_3^-(aq)$ **(4 marks)**
(b) $SO_4^{2-}(aq)$ and $CO_3^{2-}(aq)$ **(5 marks)**
(c) $K^+(aq)$ and $Na^+(aq)$ **(2 marks)**
(d) $NH_4^+(aq)$ and $Mg^{2+}(aq)$ **(5 marks)**

The result of the test on **each ion** should be stated, and **ionic** equations should be given wherever possible.

Question 20.2

A pale blue precipitate is formed when equal volumes of aqueous X and aqueous Y are mixed. The pale blue precipitate dissolves when an excess of aqueous Y is added to the mixture. X and Y could be

X	Y
A ammonia	copper(II) sulphate
B ammonia	iron(II) sulphate
C copper(II) sulphate	ammonia
D copper(II) sulphate	sodium hydroxide
E iron(II) sulphate	sodium hydroxide

(AEB, 1983)

Question 20.3

Which of the following metals forms a hydroxide which is insoluble in water but soluble in aqueous sulphuric acid or sodium hydroxide?

A aluminium
B calcium
C copper
D lead
E magnesium

Question 20.4

Sodium hydroxide solution produces a reddish-brown precipitate when added to a solution of

A aluminium sulphate
B copper(II) sulphate
C iron(II) sulphate
D iron(III) sulphate
E magnesium sulphate

Question 20.5

When solid X is heated, it gives off a brown gas. X could be
A Ammonium chloride
B Lead(II) nitrate
C Lead(IV) oxide
D Potassium nitrate
E Sodium bromide

Question 20.6

The gas evolved when concentrated sulphuric acid is added to ammonium chloride will
A bleach moist litmus paper
B form a white solid near the top of the test tube
C form steamy fumes when it comes into contact with moist air
D turn lime water milky
E turn moist red litmus paper blue.

Question 20.7

A *white* precipitate is formed when solutions of X and Y are mixed. X and Y could be
A copper(II) sulphate and sodium hydroxide
B copper(II) sulphate and barium chloride
C sodium chloride and potassium nitrate
D iron(II) chloride and sodium hydroxide
E iron(III) chloride and sodium hydroxide

Question 20.8

Which of the following procedures is NOT correct when preparing to titrate sulphuric acid solution into sodium hydroxide solution?
A Rinse the burette with the acid and then fill it with the acid.
B Run a few cm^3 of the acid out through the burette jet.
C Rinse the titration flask with alkali and then add exactly 25 cm^3 of the alkali solution.
D Put a few drops of indicator in the titration flask.
E Read the level at which the acid stands in the burette.   (NISEC)

20.6 Answers to Self-test Questions

20.1 (a) Add chlorine water. Reddish-brown bromine will be displaced from the Br^-(aq) but no change will occur in the NO_3^-(aq).

$$2Br^-(aq) + Cl_2(aq) \rightarrow Br_2(aq) + 2Cl^-(aq)$$

Test 7 could be used here.

(b) Add *cold* dilute hydrochloric acid. CO_3^{2-} (aq) will give an effervescence of carbon dioxide (turns lime water milky) but SO_4^{2-} (aq) will not react.

$$2H^+ (aq) + CO_3^{2-} (aq) \rightarrow H_2O (l) + CO_2 (g)$$

See Test 3.

(c) Carry out a flame test. K^+ (aq) gives a lilac flame and Na^+ (aq) a golden yellow one.

See Test 9.

(d) Add sodium hydroxide solution. Mg^{2+} (aq) will give a white precipitate of magnesium hydroxide.

$$Mg^{2+} (aq) + 2OH^- (aq) \rightarrow Mg(OH)_2 (s)$$

From the NH_4^+ (aq) a smell of ammonia (turns moist red litmus paper blue) may be detected in the cold. On warming, more ammonia will be evolved.

$$NH_4^+ (aq) + OH^- (aq) \rightarrow NH_3 (g) + H_2O (l)$$

See Test 7.

20.2 **C.**

See Test 8.

20.3 **A.**

Aluminium hydroxide is amphoteric; lead(II) hydroxide is amphoteric but does not dissolve in dilute sulphuric acid because lead(II) sulphate is insoluble in water.

20.4 **D**

See Test 7.

20.5 **B.**

See Test 2.

20.6 **C.**

See Test 4.

20.7 **B.**

See Test 6.

20.8 C.

The flask should be rinsed with distilled water, not alkali.

Table of Relative Atomic Masses

Element	Symbol	Atomic number	Relative atomic mass	Element	Symbol	Atomic number	Relative atomic mass
Actinium	Ac	89	227	Holmium	Ho	67	165
Aluminium	Al	13	27	Hydrogen	H	1	1
Americium	Am	95	243	Indium	In	49	115
Antimony	Sb	51	122	Iodine	I	53	127
Argon	Ar	18	40	Iridium	Ir	77	192
Arsenic	As	33	75	Iron	Fe	26	56
Astatine	At	85	210	Krypton	Kr	36	84
Barium	Ba	56	137	Lanthanum	La	57	139
Berkelium	Bk	97	249	Lawrencium	Lr	103	257
Beryllium	Be	4	9	Lead	Pb	82	207
Bismuth	Bi	83	209	Lithium	Li	3	7
Boron	B	5	11	Lutetium	Lu	71	175
Bromine	Br	35	80	Magnesium	Mg	12	24
Cadmium	Cd	48	112	Manganese	Mn	25	55
Caesium	Cs	55	133	Mendelevium	Md	101	256
Calcium	Ca	20	40	Mercury	Hg	80	201
Californium	Cf	98	251	Molybdenum	Mo	42	96
Carbon	C	6	12	Neodymium	Nd	60	144
Cerium	Ce	58	140	Neon	Ne	10	20
Chlorine	Cl	17	35.5	Neptunium	Np	93	237
Chromium	Cr	24	52	Nickel	Ni	28	59
Cobalt	Co	27	59	Niobium	Nb	41	93
Copper	Cu	29	63.5	Nitrogen	N	7	14
Curium	Cm	96	247	Nobelium	No	102	254
Dysprosium	Dy	66	162.5	Osmium	Os	76	190
Einsteinium	Es	99	254	Oxygen	O	8	16
Erbium	Er	68	167	Palladium	Pd	46	106
Europium	Eu	63	152	Phosphorus	P	15	31
Fermium	Fm	100	253	Platinum	Pt	78	195
Fluorine	F	9	19	Plutonium	Pu	94	242
Francium	Fr	87	223	Polonium	Po	84	210
Gadolinium	Gd	64	157	Potassium	K	19	39
Gallium	Ga	31	70	Praseodymium	Pr	59	141
Germanium	Ge	32	73	Promethium	Pm	61	147
Gold	Au	79	197	Protactinium	Pa	91	231
Hafnium	Hf	72	178.5	Radium	Ra	88	226
Helium	He	2	4	Radon	Rn	86	222

Rhenium	Re	75	186	Terbium	Tb	65	159
Rhodium	Rh	45	103	Thallium	Tl	81	204
Rubidium	Rb	37	85.5	Thorium	Th	90	232
Ruthenium	Ru	44	101	Thulium	Tm	69	169
Samarium	Sm	62	150	Tin	Sn	50	119
Scandium	Sc	21	45	Titanium	Ti	22	48
Selenium	Se	34	79	Tungsten	W	74	184
Silicon	Si	14	28	Uranium	U	92	238
Silver	Ag	47	108	Vanadium	V	23	51
Sodium	Na	11	23	Xenon	Xe	54	131
Strontium	Sr	38	88	Ytterbium	Yb	70	173
Sulphur	S	16	32	Yttrium	Y	39	89
Tantalum	Ta	73	181	Zinc	Zn	30	65
Technetium	Tc	43	99	Zirconium	Zr	40	91
Tellurium	Te	52	128				

Note: (1) The relative atomic masses are based on a scale where the mass of $^{12}C = 12$.

(2) For many of the radioactive elements the relative atomic mass given is the mass number of the most stable or the most common isotope.

The Periodic Table

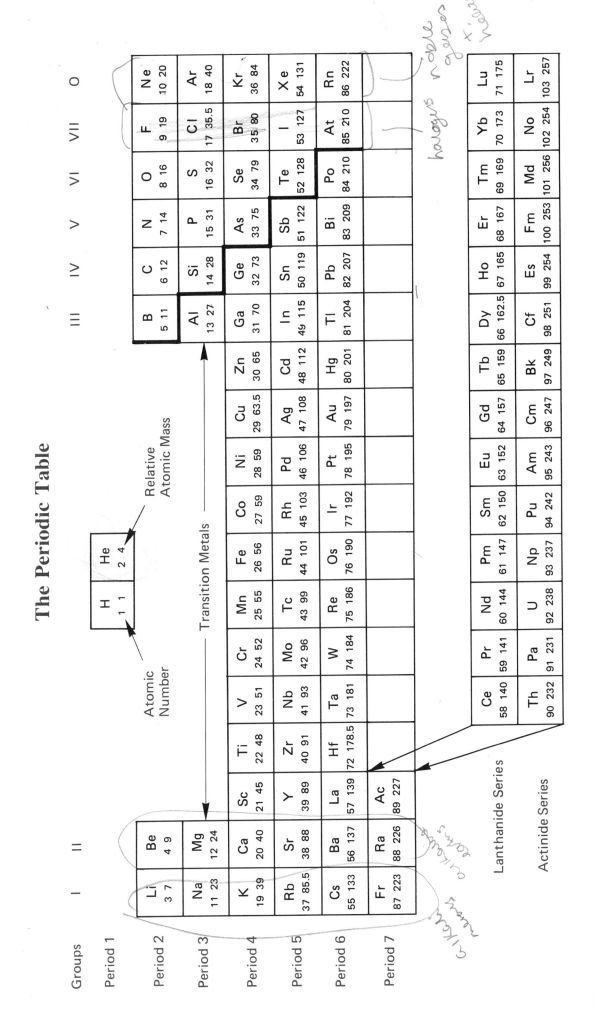

Groups	I	II													III	IV	V	VI	VII	O
Period 1	H 1 1																			He 2 4
Period 2	Li 3 7	Be 4 9													B 5 11	C 6 12	N 7 14	O 8 16	F 9 19	Ne 10 20
Period 3	Na 11 23	Mg 12 24													Al 13 27	Si 14 28	P 15 31	S 16 32	Cl 17 35.5	Ar 18 40
Period 4	K 19 39	Ca 20 40	Sc 21 45	Ti 22 48	V 23 51	Cr 24 52	Mn 25 55	Fe 26 56	Co 27 59	Ni 28 59	Cu 29 63.5	Zn 30 65			Ga 31 70	Ge 32 73	As 33 75	Se 34 79	Br 35 80	Kr 36 84
Period 5	Rb 37 85.5	Sr 38 88	Y 39 89	Zr 40 91	Nb 41 93	Mo 42 96	Tc 43 99	Ru 44 101	Rh 45 103	Pd 46 106	Ag 47 108	Cd 48 112			In 49 115	Sn 50 119	Sb 51 122	Te 52 128	I 53 127	Xe 54 131
Period 6	Cs 55 133	Ba 56 137	La 57 139	Hf 72 178.5	Ta 73 181	W 74 184	Re 75 186	Os 76 190	Ir 77 192	Pt 78 195	Au 79 197	Hg 80 201			Tl 81 204	Pb 82 207	Bi 83 209	Po 84 210	At 85 210	Rn 86 222
Period 7	Fr 87 223	Ra 88 226	Ac 89 227																	

Relative Atomic Mass
Atomic Number
Transition Metals

Lanthanide Series

Ce 58 140	Pr 59 141	Nd 60 144	Pm 61 147	Sm 62 150	Eu 63 152	Gd 64 157	Tb 65 159	Dy 66 162.5	Ho 67 165	Er 68 167	Tm 69 169	Yb 70 173	Lu 71 175

Actinide Series

Th 90 232	Pa 91 231	U 92 238	Np 93 237	Pu 94 242	Am 95 243	Cm 96 247	Bk 97 249	Cf 98 251	Es 99 254	Fm 100 253	Md 101 256	No 102 254	Lr 103 257

211

Index